Mobile Communication in Asian Society and Culture

Xie and Chao present a collection of research on mobile communication in Asian communities and countries such as Bangladesh, China, India, Japan, and South Korea. With chapters written by scholars from diverse cultural and institutional backgrounds, this book provides both localized and comparative perspectives on mobile communication research.

Exploring the way mobile apps are used in daily life in Asian countries, Xie, Chao, and their contributors analyze how mobile apps improve lives, help people build relationships, sustain communities, and change society for the better. They look at areas including the role of mobile apps in public service delivery and access, family communication, cultural norms and identities, organizational communication, and intercultural communication. The investigation of these topics elevates the understanding of the cultural, familial, interpersonal, organizational, and intercultural consequences of mobile communication in a global context. Through examining mobile apps use in regard to scale, scope, depth, complexity, and distinctiveness within the Asian context, this book furthers the research agenda of mobile communication and enriches our understanding of current practice and future direction of mobile communication.

This book serves as a useful reference for scholars and students interested to learn more of the development and application of mobile communication from a global and comparative perspective.

Ming Xie is Assistant Professor at West Texas A&M University. Ming earned her PhD in Public Administration from the University of Nebraska at Omaha and another PhD in Cultural Anthropology from the Chinese Academy of Social Sciences. She is particularly interested in nonprofit organization management and communication and emergency management.

Chin-Chung Chao is Professor of Communication at University of Nebraska at Omaha. Her primary research interests span conflict management, leadership, intercultural communication, organizational communication, and media communication. She has written two published books, twenty peer-reviewed articles, and ten book chapters. Her research has won multiple awards.

Mobile Communication in Asian Society and Culture

Continuity and Changes across Private, Organizational, and Public Spheres

Edited by Ming Xie and Chin-Chung Chao

Routledge
Taylor & Francis Group
LONDON AND NEW YORK

First published 2024
by Routledge
4 Park Square, Milton Park, Abingdon, Oxon, OX14 4RN

and by Routledge
605 Third Avenue, New York, NY 10158

Routledge is an imprint of the Taylor & Francis Group, an informa business

British Library Cataloguing-in-Publication Data
A catalogue record for this book is available from the British Library

Library of Congress Cataloging-in-Publication Data
Names: Xie, Ming, 1980– editor. | Chao, Chin-Chung, editor.
Title: Mobile communication in Asian society and culture : continuity and changes across private, organizational, and public spheres / [edited by] Ming Xie, Chin-Chung Chao.
Description: First edition. | Abingdon, Oxon ; New York, NY : Routledge, 2023. | Includes bibliographical references. |
Summary: "Xie and Chao present a collection of research on mobile communication in Asian communities and countries such as Bangladesh, China, India, Japan, and South Korea. With chapters written by scholars from diverse cultural and institutional backgrounds, this book provides both localized and comparative perspectives on mobile communication research. Exploring the way mobile apps are used in daily life in Asian countries, Xie, Chao and their contributors analyse how they improve lives, help people build relationships, sustain communities, and change society for the better. They look at areas including the role of mobile apps in public service delivery and access, family communication, cultural norms and identities, organizational communication, and intercultural communication. The investigation of these topics elevates the understanding of the cultural, familial, interpersonal, organizational, and intercultural consequences of mobile communication in a global context. Through examining mobile apps use in regard to its scale, scope, depth, complexity, and distinctiveness within the Asian context, this book furthers the research agenda of mobile communication and enriches our understanding of current practice and future direction of mobile communication. A useful reference for scholars and students interested to learn more of the development and application of mobile communication from a global and comparative perspective"– Provided by publisher.
Identifiers: LCCN 2023018930 (print) | LCCN 2023018931 (ebook) | ISBN 9781032354675 (hardback) | ISBN 9781032358369 (paperback) | ISBN 9781003328896 (ebook)
Subjects: LCSH: Information society–Asia. | Mobile communication systems–Social aspects–Asia. | Social media and society–Asia. | Electronic commerce–Social aspects–Asia.
Classification: LCC HM851 .M627 2023 (print) | LCC HM851 (ebook) | DDC 303.48/33095–dc23/eng/20230503
LC record available at https://lccn.loc.gov/2023018930
LC ebook record available at https://lccn.loc.gov/2023018931

ISBN: 9781032354675 (hbk)
ISBN: 9781032358369 (pbk)
ISBN: 9781003328896 (ebk)

DOI: 10.4324/9781003328896

Typeset in Times New Roman
by Newgen Publishing UK

Contents

Figures

Tables

Contributors

Josh Averbeck (PhD, University of Oklahoma) is Professor in the Department of Communication at Western Illinois University, where he directs the Social Media Lab. His research on language use in message design has been published in journals such as *Communication Monographs*, *Journal of Communication*, and *Human Communication Research*.

Chin-Chung Chao is Professor of Communication at the University of Nebraska, Omaha. Her primary research interests span conflict management, leadership, intercultural communication, organizational communication, and media communication. She served as the President of Association for Chinese Communication Studies (ACCS) in 2011–2012 and the Chair of Asian/Pacific American Communication Studies Division (APACS) in 2012–2013, and served as Guest Editor of the *Chinese Media Research* and *Negotiation and Conflict Management*. In addition, she has served on the editorial boards of the *Journal of Intercultural Communication Research*, *Journal of International and Intercultural Communication*, and *Negotiation and Conflict Management Research*. She has published 2 books, 20 peer-reviewed articles, and 10 book chapters. Her research has won multiple awards.

Li Chen (PhD, University of Iowa, 2016) is an associate professor at the Department of Communication in West Texas A&M University, Canyon, Texas. Chen's research focuses on health communication, in particular, the impact of health misinformation on decision-making, social media use and perceptions of health, and media representations of public health controversies.

Ye Chen (MA, University of Connecticut, 2018) is ABD in the Department of Communication at the University of Connecticut. Her research interests include the psychological effects and behavioral implications of new communication technologies, such as social media, in terms of self-disclosure, relationship maintenance, information seeking and learning, civic engagement, and well-being.

Simu Dey (MA, Western Illinois University) is a doctoral candidate in the Lew Klein College of Media and Communication at Temple University. Her research focuses on using various social media platform communication tools as channels for providing mental health support to different at-risk populations. She holds an MSS (summa cum laude) in Communication and Journalism from the University of Chittagong. She is also an Assistant Professor of Mass Communication and Journalism at Comilla University (currently on study leave). Prior to her academic pursuits, Simu worked as a general assignment reporter at the Daily Samakal, one of Bangladesh's most widely circulated Bengali newspapers.

Minshuai Ding is currently a research project manager of the Bureau of Sociological Research the University of Nebraska, Lincoln. He received his PhD in Public Administration from the University of Nebraska, Omaha. His fields of interest include public opinion research, local government, and communication in the public sector.

Lei Guo (PhD, University of Missouri) is Assistant Professor at the School of Communication, University of Nebraska, Omaha. Her research interests include the intersection of media effect, journalism studies, and media sociology.

Louisa Ha is Professor of Research Excellence (2021–24) in the School of Media and Communication at Bowling Green State University, Ohio. She is the founding Editor-in-Chief of Online Media and Global Communication, and also the former editor of *Journalism and Mass Communication Quarterly*, the flagship journal of the Association for Education in Journalism and Mass Communication. She is the founder and chair of the Emerging Media Research Cluster in the School of Media and Communication. She is the Advertising Major Advisor and Internship Coordinator. She is also the Graduate Coordinator of the School of Media and Communication's graduate programs. Her research interests are misinformation, social media influencers, online videos, new media business models, social and mobile media use, international conflicts and comparative communication, online and international advertising, and audience research.

Min Wha Han (PhD, Ohio University) is Assistant Professor in the Department of Communication at West Texas A&M University. Dr. Han specializes in rhetorical criticism, qualitative communication and media research in the East Asian context. Dr. Han has published in various peer-reviewed journals, such as *Communication Theory*, the *International Communication Gazette*, the *Review of Communication*, *Keio Communication Review*, *Language and Intercultural Communication*. Dr. Han's recent publications include a co-edited volume, "Korean Diaspora across the World: Homeland in History, Memory, Imagination, Media, and Reality" (Lexington Books, 2020), a peer-reviewed journal article, "Globalization from above and below: Rejecting superficial multiculturalism and igniting anti-Korean sentiments in Japan" (*Journal of International and Intercultural Communication*), and a book chapter, "Reframing the difference of co-ethnic other in Japan: An analysis of representations and identifications in a South Korean documentary film 'Uri-Hakkyo'" (University of Michigan Press, 2022).

Samantha James is a doctoral candidate in Organizational Communication and Technology at the University of Texas at Austin. Her research focuses on the intersection of affect, digital technology-mediated communication, and globalization. She has publications in outlets such as Seoul National University Press and has presented award-winning work at national and international conferences for communication scholarship. She is a 2022–2023 Fulbright Research Award recipient.

Md Enamul Kabir is a doctoral candidate at Bowling Green State University in the United States. Kabir studies the use of computational approaches in communication, with a focus on machine learning applications. His research centers around communication and racism, social media activism, misinformation, and pedagogy. His recent award-winning research created a novel scale to gauge "the degree of racial identity" in modern American classrooms.

Kenneth Lachlan (Ph.D., Michigan State University, 2003) is Professor and Department Head in the Department of Communication at the University of Connecticut. His current research interests include the functions and effects of social media during crises and disasters, and the use of social robotics in delivering risk messages. Lachlan is ranked among one of the top 1% of published scholars in key journals in Communication between 2007 and 2011, and between 2012 and 2016. He is a former editor-in-chief of *Communication Studies* and a coauthor of several textbooks, including *Risk and Crisis Communication: Communicating in a Disruptive Age, Introduction to Computer Mediated Communication: A Functional Approach* (1st and 2nd editions), and *Straight Talk about Communication Research Methods* (1st, 2nd, and 3rd editions).

Jeremy Harris Lipschultz (PhD, Southern Illinois University-Carbondale) is Peter Kiewit Distinguished Professor at School of Communication, University of Nebraska at Omaha. His research focuses on social media listening and engagement, journalism, and media ethics and law.

Shudipta Sharma (MPhil, Jahangirnagar University) is a doctoral student in the School of Media and Communication at Bowling Green State University, USA. He is currently on a leave from the University of Chittagong, Bangladesh where he holds an Assistant Professor position at the Department of Communication and Journalism. His research interests include political communication, artificial intelligence, digital politics, online radicalization, digital journalism, and media sociology.

Roma Subramanian, PhD, is Assistant Professor in the School of Communication at the University of Nebraska Omaha. She is also an affiliate faculty member in the university's medical humanities program and the Goldstein Center for Human Rights. Her research focuses on health communication, particularly regarding stigmatized health issues (e.g., mental illness, sexual abuse). Other areas of interest include the impact of social and mobile media on health, the dynamics of patient-provider communication, and the intersection between art and health. Her recent work has been published in *Critical Public Health* and *the Journal of Visual Communication in Medicine*.

Ming Xie is Assistant Professor at West Texas A&M University. Ming earned her PhD in Public Administration from the University of Nebraska at Omaha, and another PhD in Cultural Anthropology from the Graduate School of Chinese Academy of Social Sciences in Beijing. She is particularly interested in non-profit organization management and communication, social media, emergency management, and intercultural communication.

Dongdong Yang (MA, Wake Forest University, 2018) is ABD in the Department of Communication at the University of Connecticut. Yang is primarily interested in how cross-cultural differences between China and the United States influence the use and effects of new communication technologies, such as social media. She also examines how identity influences perceptions and attitudes in sports communication. Yang is a recipient of three top paper awards at the AEJMC annual conferences.

Ping Yang (PhD, Arizona State University) is Associate Professor in the Department of Communication and Theatre at Millersville University, Pennsylvania. She holds her Doctorate in Intercultural Communication, an MA in Communication, a second MA in Linguistics, and a BA in English Language and Culture. Her research and teaching interests are located at the intersection of culture, communication, and technology. Her publications focus on identity negotiation, cultural adaptation, media representation, computer-mediated communication, and intercultural communication competence.

Yafei Zhang is Assistant Professor in the School of Journalism and Communication at the Renmin University of China. Zhang's research focuses on strategic communication, corporate social responsibility, big data analysis, and social network analysis.

Ge Zhu, PhD, is Research Associate at the College of Public Health, University of Iowa, Iowa City, Iowa. Her research contributes to the area of health communication and utilizes multiple methodologies, including conventional quantitative research methods, computer-assisted text mining, and social network analysis. Ge's current work focuses on social media campaigns addressing vaccine hesitancy in U.S. Midwest rural and micropolitan communities.

Acknowledgments

We extend our greatest appreciation to many people who helped to make this book a reality. First and foremost, thanks to the authors of the chapters contained within these pages and the peer reviewers who provided extremely helpful feedback.

We are grateful to our colleagues for the inspiration and ideas that led us to work on this project. This book combines three refereed articles published in a special issue on mobile apps in China in *China Media Research* with modifications. We would like to acknowledge *China Media Research*'s official permission to reprint these three articles.

Of course, this book would not have been possible without Yongling Lam (the former editor at Routledge), Kendrick Loo (current editor at Routledge), and Chelsea Low Yingqi (editorial assistant at Routledge) and their willingness to believe in and support this project. Their prompt responses, continuing support, and expert advice helped keep the book on the right track and ready for publication in a timely manner.

Last but not least, thank you to our family members for their unfailing and loving support without which this endeavor would not have been possible.

Ming Xie and Chin-Chung Chao

1 Introduction to Mobile Communication in Asian Society

Ming Xie

Introduction

Mobile communications are becoming mainstream communications and provide opportunities for human and social development. According to Statista, there were more than six billion smartphone subscriptions worldwide as of 2021 (Taylor, 2023). China, India, and the United States are the countries with the highest number of smartphone users (Laricchia, 2022). Also, the World Bank's Information and Communication for Development Report (2012) states that in some developing countries, more people have access to a mobile phone than to clean water, a bank account, or even electricity. In Afghanistan, 91 percent of households reported owning a mobile phone (Lebrun et al., 2020). A person from a remote rural village in Afghanistan can connect to the world by a cell phone. If mobile apps have become a ubiquitous phenomenon all over the world, it is necessary to explore how mobile apps are embedded into an array of communication strategies and change our communicative ways in every aspect of our lives.

Mobile Communication in Asia

Asia has been seen as the epicenter of mobile communications (Rao & Mendoza, 2004; Srivastava, 2008). Japan, South Korea, India, and China have been the leading countries of 3G and 5G technology all over the world. Also, Asian countries such as China, India, Indonesia, Japan, Vietnam, and Bangladesh are among the top countries with the most mobile phone subscribers (Laricchia, 2022). In 2004, the book *Asia Unplugged: The wireless and mobile media boom in the Asia-Pacific* was published by Sage. It provides an overview of the system and of different players in mobile communications in Asian countries and highlights the unique role of Asia. After 19 years, the landscape of mobile communication has changed dramatically with the emergence of numerous mobile apps. TikTok, a short video app that originated in China, is coming to dominate social media platforms and has captured billions of users' attention around the world. During the COVID-19 pandemic, mobile apps-based education, sports, and businesses have been the only way people interact, entertain, work, and have social lives. All of these offer unique

DOI: 10.4324/9781003328896-1

opportunities for breadth and depth in academic research and comparative analysis of mobile app use and communication in the diverse region of Asia.

The complex history and geopolitics have shaped and reshaped the current community and society of Asia. Despite the rapid growth and development of information technology in the region, there is still a significant disparity regarding socioeconomic status among Asian countries. China is the second largest economy while Bangladesh is still one of the lower-middle-income countries in the world (World Bank, 2022). In terms of technology, the mobile penetration in e-commerce in Asia is 74 percent compared to 35 percent in Europe and 31 percent in North America. However, more than two-thirds of the technology hotspots are located in China and India (Swaminathan, 2022). Also, telecommunications companies and business processing centers are in India and the Philippines (Dutta & Shome, 2018). Overall, mobile communication affords connectivity and mobility across physical, geographical, and economic boundaries. Several questions emerge in thinking about mobile communication in contemporary Asian society: whether and how mobile devices and technologies promote inclusive development in the region; whether and how mobile apps increase the health literacy of individuals and change their health attitudes and behaviors; how mobile apps improve processes of governance, provide accountability, transparency, and responsiveness for the interaction between government and citizens; how mobile apps are used by political and social activists to promote social change, social movements, and innovation; and how mobile apps help overcome the culture and language barriers and cultural adjustment. These questions are not an exhaustive list regarding the application of mobile communication in our social life. However, exploring these questions in Asia does provide a contextual perspective regarding the continuity and changes in Asian culture, society, and human behaviors in the era of mobile communication.

The Aims and Objectives of the Book

This book is a collection of research on mobile communication in Asian communities and countries such as Bangladesh, China, India, Japan, and South Korea. It has three objectives. First, to offer rich empirical and contextual evidence regarding how mobile apps empower individual users, enrich lifestyles and livelihoods, and boost the economic and social development in Asian societies and cultures. Second, to contribute to the research agenda of mobile communication through the exploration of mobile app use regarding its scale, scope, depth, complexity, and distinctiveness within the Asian context. Three, to examine how Asian cultures have been changed and have adapted to the mobile communication environment. The goal of the book is to provide in-depth coverage of the diverse Asian cultures from an indigenous perspective rather than view Asian communities and cultures as monolithic from a Western viewpoint.

Organization of the Book

The book explores how mobile apps are used in various aspects of our daily life and society in Asian countries to improve lives, help people build relationships, sustain communities, and change society for the better. We invited scholars with expertise in mobile communication research from diverse cultural and institutional backgrounds to provide both localized and comparative perspectives. The topics include the role of mobile apps in public service delivery and access, family communication, cultural norms and identities, organizational communication, and intercultural communication.

Chapter 1 offers an introduction to the significance of studying mobile communication in Asian society and communities. Following this chapter, 12 studies were divided into four parts with three chapters in each: mobile communication in the private sphere, mobile communication in the organizational sphere, mobile communication in the public sphere, and mobile communication within the network society. The three chapters in Part 1 focus first in Chapter 1 on mobile communication in relation to empty nesters, gender schema, and on at-risk populations' mental health. Chapter 2 explores Chinese midlife empty-nest parents' WeChat learning, and gaming behaviors. Chapter 3 adopts gender schema theory and digital feminism framework and discusses how the role of housewives and gender issues are portrayed on social media in Chinese society. Turning to the context of Bangladesh, Chapter 4 explores at-risk populations' self-disclosure and the impact on their mental health. Part 2 includes three studies on mobile communication within the organizational sphere. Chapter 5 is a field study on mobile communication in flood emergency management in China with a focus on collaboration among public, nonprofit, and private organizations. Chapter 6 examines livestream e-commerce and the communicative strategies of the two largest Chinese e-commerce companies, Taobao and Douyin. Chapter 7 focuses on three Indian mental health non-governmental organizations' use of Instagram and discusses how these organizations deliver mental health care and address mental illness stigma by leveraging social media affordances. Focusing on mobile communication in the public sphere, the next three chapters cover studies on the role of mobile communication in health promotion in India and China and in social movements in Bangladesh. Chapter 8 compares users' behaviors on different smartphone operating systems in Indian health promotion on Twitter. Chapter 9 investigates WeChat users' health information-seeking activities during the COVID-19 pandemic in China. Focusing on the context of Bangladesh with low socioeconomic status and limited access to information and communication technologies, Chapter 10 explores how mobile communication can facilitate the transition of an online social movement to an offline one. The discussions of the three chapters in Part 4 are around the application of mobile communication in the contemporary network society. Chapter 11 adopts a social network analysis to explore how Twitter users produce counter-speech to fight against anti-Asian hate and crimes. From a globalized perspective, Chapter 12 explores Japanese immigrant housewives in the United States and their use of LINE to maintain friendships and cultural identities. In contrast, the author of Chapter 13 is an American fan of Korean popular culture. Her study provides a vivid picture regarding the role of mobile communication in helping her navigate Korean fan culture and society. Finally, the book concludes with a chapter that reviews the lessons learned from the 12 studies in the book and discusses possible future directions of research on mobile communication.

The exploration of the above topics is helpful in understanding the cultural, familial, interpersonal, organizational, and intercultural consequences of mobile communication in a global context. The exploration of mobile app use about its scale, scope, depth, complexity, and distinctiveness within the Asian context will contribute to the research agenda of mobile communication and enrich our understanding of the current practice and future direction of mobile communication.

References

Dutta, M. J., & Shome, R. (2018). Mobilities, communication, and Asia. *International Journal of Communication, 12*, 3960–3978.

Laricchia, F. (2022, October 18). *Smartphone users by country 2021 | Statista*. www.statista.com/statistics/748053/worldwide-top-countries-smartphone-users/

Lebrun, V., Dulli, L., Alami, S. O., Sidiqi, A., Sultani, A. S., Rastagar, S. H., Halimzai, I., Ahmadzai, S., & Todd, C. S. (2020). Feasibility and acceptability of an adapted mobile phone message program and changes in maternal and newborn health knowledge in four provinces of Afghanistan: Single-group pre-post assessment study. *JMIR MHealth and UHealth, 8*(7), e17535. https://doi.org/10.2196/17535

Rao, M., & Mendoza, L. (2004). *Asia Unplugged: The wireless and mobile media boom in the Asia-Pacific*. SAGE Publications. http://ebookcentral.proquest.com/lib/unomaha/detail.action?docID=434182

Srivastava, L. (2008). The mobile makes its mark. In J. E. Katz (Ed.), *Handbook of mobile communication studies* (pp. 15–28). The MIT Press.

Swaminathan, A. (2022, March 15). *How can Asia narrow the gap between countries' access to technologies, and leapfrog to sustainable, inclusive growth?* McKinsey. www.mckinsey.com/featured-insights/future-of-asia/videos/how-can-asia-narrow-the-gap-between-countries-access-to-technologies-and-leapfrog-to-sustainable-inclusive-growth

Taylor, P. (2023, January 18). *Smartphone subscriptions worldwide 2027*. Statista. www.statista.com/statistics/330695/number-of-smartphone-users-worldwide/

World Bank. (2022, October 6). *The World Bank in Bangladesh* [Text/HTML]. World Bank. www.worldbank.org/en/country/bangladesh/overview

Part I

Mobile Communication in the Private Sphere

2 WeChat Gaming, Learning, and Midlife Empty Nest

Dongdong Yang, Kenneth Lachlan, and Ye Chen

Introduction

Over the past 40 years, China's reform and opening-up policy has given rise to a modern society featuring physical mobility (Lerner, 1958). In 2011, China recorded over 254 million internal migrants, accounting for 20 percent of the total population (China's National Bureau of Statistics, 2012). When millions of young people leave their hometowns for academic or professional development opportunities, their midlife parents become empty nesters. In this context, how midlife parents regulate negative emotions (e.g., loneliness) commands academic attention.

The Internet has been widely adopted as an emotion-regulation strategy (Wolfers & Schneider, 2021). However, middle-aged citizens' social media use patterns are understudied, despite their stronger financial and intellectual capacities for new media consumption (Huang & Zhang, 2017). Limited research suggests that playing digital games reduces older adults' loneliness (Schell et al., 2016), and participation in non-formal learning also improves their psychological well-being (Narushima et al., 2016). The current study thus focuses on the gaming and learning activities Chinese midlife parents engage with on WeChat: an all-encompassing social media platform that has successfully penetrated the middle-aged demographic (Huang & Zhang, 2017).

Drawing from uses and gratifications (Katz et al., 1973) and self-determination (Deci & Ryan, 1984) theories—which posit that media use is motivated by psychological needs, and the satisfaction of such needs enhances intrinsic motivation—the current study employs a survey design to examine whether children's residential status predicts midlife parents' loneliness, alongside how loneliness affects gaming and learning behaviors on WeChat. Specifically, for the relationship between loneliness and WeChat learning, motivation to learn is proposed as a mediator, and WeChat proficiency as a moderator. Consistent with Chen et al. (2023), which illustrates how using information and communication technologies (ICTs) helps promote successful aging, this study provides quantitative support for the role of WeChat in satisfying midlife parents' psychological needs.

DOI: 10.4324/9781003328896-3

Empty Nest

Midlife is an age range on which academia has not reached a consensus, but the lower limit should be at least 35 (Alleyne, 2010), with the upper limit being 65 (Quinn, 2014). During this period, people experience changing roles and family events (Schafer & Shippee, 2010), including children's home leaving, following which they enter a new phase of the adult life cycle, coined as "the empty nest" (Harkins, 1978). In the Chinese context, empty nest feelings are mostly negative, as nest leaving may be reflective of parental failure in instilling family values (e.g., filial piety) in children (Goldscheider & Goldscheider, 1999), which causes loneliness.

Loneliness is a negative emotional state experienced by empty nesters when the quality of relationships they perceive to have cannot meet their perceived relational needs (Alberti, 2019). For instance, Wang et al. (2017) found significant differences in loneliness between empty-nest and non-empty-nest elders in rural China. Because of China's one-child policy that was implemented from 1979 to 2015, Chinese parents of only children on average become empty nesters three years earlier than those who have multiple children, with fathers and mothers entering the empty nest phase at the age of 49 and 47, respectively (Feng, 2009). Moreover, increasing physical mobility and the embrace of a global perspective have motivated growing numbers of underage students to leave their parents and study abroad (e.g., Larmer, 2017). Thus, more Chinese parents are projected to become empty nesters younger than 43 years old (Zhu, 2021). In this case, midlife parents are less likely to have past experience concerning how to face their children's leaving, and they may still be too young to have lowered their expectations to cope with such negative empty-nest feelings (e.g., Mitchell & Lovegreen, 2009). Under this premise, it is suggested that:

H1a: Chinese midlife empty-nest parents will be lonelier than non-empty nesters.

Moreover, mothers have reported experiencing greater distress than fathers when children leave home, because women have typically spent more time in child rearing and built a stronger bond with their kids (Mitchell & Lovegreen, 2009). Research conducted in China also demonstrates that first-year college students' empty-nest mothers are more paranoid than fathers (Zhang, 2012). Similarly, a survey that enrolled 4,901 empty nesters in Shanxi Province of China shows that women perceived lower life satisfaction than men (Zhang et al., 2021). Therefore, it is posited that:

H1b: Gender will moderate the effect of children's residential status on the level of loneliness among midlife parents in China, with empty-nest women being lonelier than men.

Uses and Gratifications

In order to link empty nest feelings to social media use, uses and gratifications theory (U&G) is employed in the current study. U&G suggests that cognitive, emotional, social, and habitual needs lead to differential patterns of mass media exposure, which results in gratifications (Katz et al., 1973). These four categories of needs can also explain social media use (Wang et al., 2012). For instance, Dolan et al. (2016) classified social media content into informational, entertainment, remunerative, and relational content, according to the user needs that can be satisfied. Notably, remunerative needs are less relevant to the study context, while relational use of social media has been examined by prior studies (Huang & Zhang, 2017; Tanis et al., 2017). For instance, Huang and Zhang (2017) found that people aged 46–55 increased the frequency of interactions with their children and relatives, yet lowered the frequency of interacting with friends using social media than the group aged 36–45; they attributed this change to children's residential status. Based on the finding that empty-nest parents indeed change their social media routines in response to children's home leaving, the current study examines how midlife parents engage in online learning and gaming on social media, which have been understudied by prior researchers. Specifically, playing WeChat games can be seen as consumption of entertainment content driven by Chinese midlife parents' emotional needs, while learning on WeChat is considered as instrumental media use that satisfies their cognitive needs.

WeChat Gaming and U&G

Boasting 1.29 billion monthly active users in the first quarter of 2022 (Statista, 2022), WeChat has topped Weibo and become the most popular social media across demographics in China, including urban inhabitants older than 40 years old (Chinese Social Science Net, 2015). WeChat features multimedia instant messaging, social network service, mini-games, broadcasting accounts for users' subscriptions, and dozens of other civic services. Since WeChat has collapsed the boundaries of different applications, it can satisfy the entertainment and information-seeking needs of its middle-aged users (Huang & Zhang, 2017) and allow researchers to study both gaming and learning behaviors on the same platform.

U&G has been widely applied to online gaming research, including casual games that allow gamers to cope with challenging life experiences (Pallavicini et al., 2022). WeChat mini-games are built-in casual games that feature lower cognitive loads and time demands, which meet users' entertainment needs during fragmented time periods (Zhang, 2018). Although digital games have emerged as a major type of entertainment for middle-aged Chinese users (Wang, 2017), this particular group constitutes an overlooked gaming sector, as the presumption that gamers are young people is still a prevalent stereotype (De Schutter et al., 2015).

As one of the most cited media activities in terms of emotion regulation (Wolfers & Schneider, 2021), gaming may meet the needs of midlife empty nesters who frequently reported loneliness (Tanis et al., 2017). Finn and Gorr (1988) discovered that loneliness correlates positively with the social-compensation media use motives, including escaping, habit, and passing time; these motives can be well satisfied by social media use (Wang et al., 2012) and online gaming (Demetrovics et al., 2011). Although direct evidence that links loneliness to mobile game playing among Chinese midlife parents is insufficient, prior research has established the positive link between loneliness and mobile social game intensity within other demographics (e.g., Chen & Leung, 2016). According to U&G, emotional social media use is driven by the strength of emotional needs (e.g., Wang et al., 2012), so it is hypothesized:

H2: Loneliness will be positively related to game playing on WeChat.

WeChat Learning and U&G

Information seeking on social media gratifies cognitive needs and indirectly reduces older adults' loneliness (Yang et al., 2021). WeChat provides a convenient platform for midlife parents to seek information, develop their hobbies, or even advance their careers. WeChat supports educational mini-programs, which cover a broad range of subjects, including but not limited to the English language, classical Chinese poetry, and outdoor activities. Online training programs organized in WeChat groups have also attracted middle-aged people to join. Users typically pay a small amount of tuition to enter such groups in order to enjoy one-on-one interactions with the instructor, which brings people back to the mentorship education model (Chen, 2016). The current study defines WeChat learning as actively using WeChat to obtain information and instructions that promote health, personal development, and professional advancement (Petik et al., 2013). Since maintaining cognitive abilities represents a key dimension of successful aging (Rowe & Khan, 1997), it is important to examine midlife parents' learning behavior on their most frequently used social media.

As put by Mitchell and Lovegreen (2009), the empty nest provides a good opportunity for parents to "rekindle interests" (p. 1652) in Western contexts. However, lifelong learning in China often embraces a utilitarian orientation (Chen & Liu, 2019), making Chinese midlife parents less likely to engage in active learning, especially when they feel lonely. From a neurocognitive perspective, "human connections shape neural connections" (Siegel, 2001, p. 3). Specifically, loneliness deprives cognitive resources (Hawkley & Cacioppo, 2010) and discourages academic participation (Bek, 2017) while increasing difficulty in filtering out irrelevant information (Baumeister et al., 2005). Thus, we expect that lonelier midlife parents are less likely to engage in learning activities on WeChat as a result of reduced cognitive resources and a sense of purpose. More formally:

H3: Loneliness will be negatively related to learning on WeChat.

Related conceptions concerning midlife parents' learning behavior on WeChat can be further informed by a consideration of motivation to learn and technology proficiency.

MOTIVATION TO LEARN. Self-determination theory (SDT) (Deci & Ryan, 1984; Ryan & Deci, 2000) suggests that satisfaction of psychological needs for competence, autonomy, and relatedness increases intrinsic motivation that, in turn, facilitates learning in various educational contexts, including online learning. For instance, Yang et al. (2014) uncovered that learning activities that occur on social networking sites could open opportunities for self-motivated students to benefit from informal exploration.

Unlike full-time students who are frequent subjects of educational psychology and communication research, midlife parents may vary substantially in their past, present, and potential future engagements in educational activities (Gorges et al., 2016). Petik et al. (2013) demonstrated that middle-aged and elderly individuals with stronger intrinsic motivations are more active in offline self-developing learning; new life situations presented by major life events (e.g., children's leaving home) may open them up to self-development programs. When it comes to older adults' social media use, Chen et al. (2023) observed that cognitive mastery and control of ICTs—antecedents to intrinsic motivation—predict active online media use, which in turn, promotes successful aging. Therefore, for midlife parents in the current study, it is hypothesized:

H4a: Motivation to learn will be positively related to learning on WeChat.

However, learning requires a higher cognitive workload, which can hardly be achieved with negative emotions, such as loneliness. Past research supports the notion that loneliness reduces the sense of meaning in life (Mwilambwe-Tshilobo et al., 2019). Lonely people thus tend to make less self-regulation effort and quit sooner on a demanding task (Baumeister et al., 2005). According to U&G, individuals understand their emotional needs and choose corresponding media to gratify such needs. As midlife parents who feel lonelier may rely on mobile games to pass time and take their minds off loneliness, they may be less likely to have strong motivations to learn. Thus, we hypothesized that:

H4b: Loneliness will be negatively related to motivation to learn.

WECHAT PROFICIENCY AND LEARNING. Despite the great variety of affordances provided by WeChat, most users do not frequently conduct learning on the platform. This is especially true for midlife parents who are only digital immigrants and selectively adopting digital skills (Olson et al., 2011). Moreover, social media proficiency heightens cognitive mastery and control—two constructs mirroring perceived competence and autonomy in SDT—which contribute to the active use of ICTs in general (Chen et al., 2023) and WeChat learning in particular. Since proficient users expect themselves to gain more from the Internet's positive effects (Kim & Hancock, 2015), midlife parents need a high WeChat proficiency in order to participate in and benefit from the learning activities on WeChat. As social media requires users to grasp more skills to navigate online environments compared to traditional media (Lev-on, 2017), only people who are more proficient in using WeChat tend to engage in more learning activities. To verify this assumption, we postulate this hypothesis:

H5: WeChat proficiency will moderate the relationship between loneliness and WeChat learning, with highly proficient midlife parents engaging in more learning activities than less proficient parents.

In addition, we anticipate that empty nest status will directly influence Chinese midlife parents' WeChat gaming and learning activities. But given the lack of research in this area, past theory provides little guidance for further conceptualization. Therefore, the following research questions are proposed:

RQ1 (a-b): Will empty-nest parents differ in WeChat (a) gaming and (b) learning activities from non-empty nesters?

Method

Participants and Procedures

With advance IRB approval, participant recruitment information was posted on the researchers and their friends' WeChat timelines to collect data through a snowball sampling procedure. Participants were encouraged to share the link with their networks to invite more respondents. Snowball sampling is often used to survey hard-to-reach populations, such as lone mothers across age groups (Duncan & Edwards, 1999). The questionnaire was administered in Chinese. Two bilingual authors translated the instruments from English into Chinese and back into English to ensure semantic consistency. Data was collected in February 2021 when there was no major pandemic lockdown in China.

After logging onto the study website, 545 participants first answered two screening questions, respectively about their age and number of children, with 111 respondents who indicated that they were not aged 35–65 or without a child being filtered out. Participants then rated their WeChat learning and gaming intensity, WeChat use proficiency, motivation to learn, and loneliness using seven-point Likert-type scales (see Table 2.1). Respondents who failed the attention check were excluded from later analysis, yielding 276 (N_{women} = 173) usable responses from midlife parents aged 35–62 living across 20 provinces of China. The sample included 103 (37.3%) parents who were empty nesters at the time of data collection; 58 (21%) of the participants were under 40 years old, 81 (29.3%) aged 40–49, 112 (40.6%) aged 50–59, and 10 (3.6%) aged 60–62. Empty-nest parents ($M = 52.72$, $SD = 6.36$) were older than those who had children at home ($M = 43.96$, $SD = 6.84$, $t(204.55) = 10.41$, $p < .001$). Additionally, 235 respondents (83.9%) had only one child, and 44 (15.7%) had two children, while only one person (0.4%) reported having three or more children. The sample size met the threshold suggested by Kline (2015), such that the minimum sample size should exceed 10 times of the total number of parameters to be estimated in a path analysis.

Table 2.1 Measures

Variable	*Item*
WeChat Learning (微信学习) *M* = 2.82, *SD* = 1.22, *Median* = 2.60, *Max* = 7.00, *Min* = 1.00	How much time on an average day do you spend on WeChat (平均每天你花多长时间上微信): 1) Learning foreign languages 学习外语 2) Learning professional skills 学习专业知识 3) Learning about health issues 了解健康问题 4) Learning something related to your hobbies 发展兴趣爱好 5) Learning about other self-development aspects 为自身发展而学习
WeChat Gaming (微信游戏) *M* = 1.50, *SD* = .82, *Median* = 1.00, *Max* = 6.33, *Min* = 1.00	How much time on average do you spend daily playing games on WeChat? 平均每天你花多长时间玩微信小游戏?
WeChat Proficiency (微信熟练程度) *M* = 4.53, *SD* = 1.07, *Median* = 5.00, *Max* = 6.00, *Min* = 1.00	From (1) not at all proficient to (7) very proficient, what is your level of WeChat proficiency? 用从"（1）完全不熟练"到"（7）非常熟练"的标准衡量你的微信熟练程度， 你认为自己处于什么水平?
Motivation to Learn (学习动机) *M* = 5.18, *SD* = 1.06, *Median* = 5.25, *Max* = 7.00, *Min* = 1.00	I like learning new things. 我喜欢学习新东西。 I like to get to the bottom of difficult things. 遇到难题时，我喜欢刨根问底。 I like to figure out how different ideas fit together. 我喜欢探究不同观点的共通之处。 If I don't understand something, I look for additional information to make it clearer. 遇到难理解的问题时，我会搜集额外信息以探究竟。
Loneliness (孤独) *M* = 2.03, *SD* = .80, *Median* = 2.00, *Max* = 4.89, *Min* = 1.00	I feel alone when I'm with my family. 当我和家人在一起的时候，我感到孤独。 No one in my family really cares about me. 家里没人关心我。 There is no one in my family I can depend upon for support and encouragement, but I wish there were. 我家里没人能给我所期望的支持和鼓励。 I really care about my family. 我非常在乎我的家人。 I really belong in my family. 我对家庭很有归属感。 I feel part of my family. 我感觉自己是家庭中的重要一员。 My family really cares about me. 我的家人非常关心我。 My family is important to me. 家庭对我很重要。 I feel close to my family. 我感到和家人很亲近。

Source: Table created by authors during analysis.

Measures

To measure **WeChat learning**, the participants evaluated how much time on an average day they would spend on WeChat, learning foreign languages, professional skills, health issues, something related to their hobbies, and other self-development areas, adapted from Petik et al. (2013) (M = 2.82, SD = 1.22, α = .75). **WeChat gaming** was measured with a single item inquiring how much time, on average, people spent daily playing games on WeChat (M = 1.50, SD = .82). **WeChat proficiency** was measured with the item adapted from Lev-on (2017) (M = 4.53, SD = 1.07). Allen et al. (2022) noted that most single-item measures prove to be as valid and reliable as their multi-item counterparts, especially when a construct is unambiguous. For **motivation to learn**, four items from Gorges et al. (2016) were adopted (M = 5.18, SD = 1.06, α = .86). Instead of the revised UCLA loneliness scale that treats loneliness as a unidimensional construct, **loneliness** in this study was measured with the Family Loneliness sub-scale from the Social and Emotional Loneliness Scale for Adults (SELSA) (DiTommaso & Spinner, 1993) (M = 2.03, SD = .80, α = .85). Detailed information about each measure is presented in Table 2.1. Moreover, for children's **residential status**, respondents indicated whether their only child/youngest child was still living with them. For families with multiple children, only when the youngest child had left home would the parents be considered empty nesters (empty nest = 1, non-empty nest = 0). Information on **age** (M = 47.21, SD = 7.91), **number of children they have** (M = 1.16, SD = .38), and **daily average work time** (M = 7.68, SD = 3.18) was also collected as **control variables**, because these three variables have been demonstrated to influence mastery of ICTs, successful aging (Chen et al., 2023), and how well parents cope with negative empty-nest emotions (Mitchell & Lovegreen, 2009; Xu & Yang, 2023; Zhu, 2021).

Results

Descriptive statistics for the variables and their zero-order correlations are reported in Table 2.2. The highest correlation was found between WeChat proficiency and motivation to learn (r = .31, p < .01). Hence, there was no multicollinearity concern.

A two-step procedure was employed to conduct the structural equation modeling (SEM) analysis to test all the hypotheses and research questions (Kline, 2015). First, the measurement model demonstrated sufficient fit: χ^2 = 218.21, df = 154, $PCMIN/df$ = 1.41; CFI = .968, $RMSEA$ = .039 (Hu & Bentler, 1999). Second, measurement invariance was tested between men and women. Results showed that partial scalar invariance could be assumed. This helped lay the groundwork for testing H1b in the following sections.

Next, a path analysis was performed using MPlus 8, which demonstrated a good model fit: χ^2 = 1.39, df = 3, p = .71; CFI = 1.000, $RMSEA$ = .000 (see Figure 2.1). H1a anticipated that midlife empty-nest parents were lonelier than non-empty nesters. The estimation results confirmed this hypothesis (β = .27, p < .001). H2 expected that loneliness would increase game playing, which was verified (β = .15,

Table 2.2 Zero-Order Correlations and Descriptive Statistics

	1	*2*	*3*	*4*	*5*	*6*
1 Empty Nest (0=No, 1=Yes)	-					
2 Loneliness	−.03	-				
3 Motivation to Learn	−.07	−.22**	-			
4 WeChat Learning	.01	.02	.22**	-		
5 WeChat Gaming	.05	.07	.03	.21**	-	
6 WeChat Proficiency	−.22**	−.09	.31**	.12*	.05	-
M		2.03	5.18	2.82	1.50	4.53
SD		.80	1.06	1.22	.82	1.07
*p*** < .01; *p** < .05						

Source: Table created by authors during analysis.

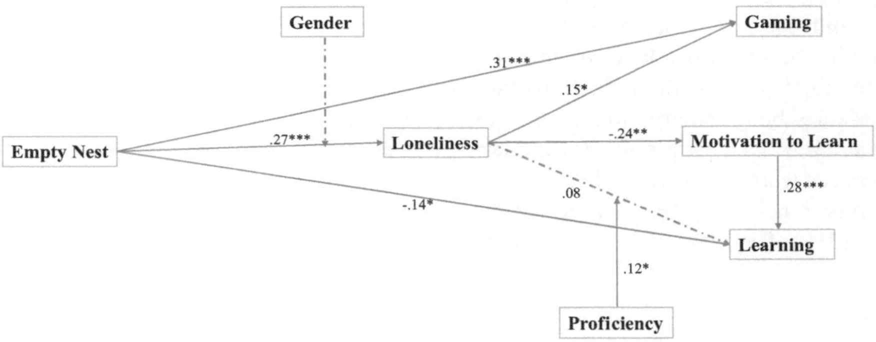

Figure 2.1 Model Results.

$p = .016$). However, contrary to our prediction, the relationship between loneliness and WeChat learning ($\beta = .08$, $p = .210$) was not significant, failing to support H3. For H4a, results rendered support to the hypothesis that motivation to learn would be positively associated with WeChat learning activities ($\beta = .28$, $p < .001$). H4b posited that loneliness would reduce motivation to learn, which received support ($\beta = -.24$, $p < .001$). Moreover, we examined whether midlife empty nesters and non-empty nesters differed in WeChat gaming (RQ1a) and learning (RQ1b) activities. Results suggested that empty-nest parents engaged in significantly more gaming ($\beta = .31$, $p < .001$) and fewer learning activities ($\beta = -.14$, $p = .025$) on WeChat than their counterparts. The results are presented in Figure 2.1.

Bootstrapping estimation was used to examine the indirect effects (Preacher & Hayes, 2008) in the model with 5,000 bootstrap samples and 95 percent bias-corrected confidence intervals (CIs). The results demonstrated that, controlled for the direct effect, the indirect effect of empty nest on WeChat gaming via loneliness was significant ($\beta = .04$, 95% CI [.003, .12]). In contrast, the indirect effect of empty nest on WeChat learning via loneliness was not significant ($\beta = .003$, 95%

CI [−.04, .05]). However, through the path that includes both loneliness and motivation to learn as mediators, empty nest had a significantly negative indirect effect on WeChat learning (β = −.02, 95% CI [−.04, −.01]). Loneliness also emerged as a significant mediator between empty nest and motivation to learn (β = −.07, 95% CI [−.13, −.02]). Motivation to learn mediated the relationship between loneliness on WeChat learning (β = −.07, 95% CI [−.13, −.03]).

In order to test whether gender moderates the relationship between empty nest and loneliness, a multiple-group analysis was conducted. As previously stated, partial scaler equivalence was established between men and women. We compared the model with no constrained paths to the model that only constrained the path between loneliness and empty nest to be equal across the two groups. Since Δdf = 1, *CMIN* = .311, p = .58, the two nested models were not significantly different. Therefore, gender did not significantly affect the relationship between empty nest and loneliness. Thus, H1b was not supported.

Finally, Hayes's (2017) PROCESS procedure (Model 5)—with loneliness being the independent variable, motivation to learn as the mediator, WeChat learning serving as the dependent variable, WeChat proficiency acting as the moderator, age, number of children, daily average work time, and all other above-mentioned variables being control variables—yielded significant results (β = .12, p = .048). Therefore, WeChat proficiency moderates the relationship between loneliness and WeChat learning. Specifically, when proficiency was at high (versus middle or low) levels, loneliness was positively related to WeChat learning (β =.19, *SE* = .08, p = .015). The interaction effect is presented in Figure 2.2.

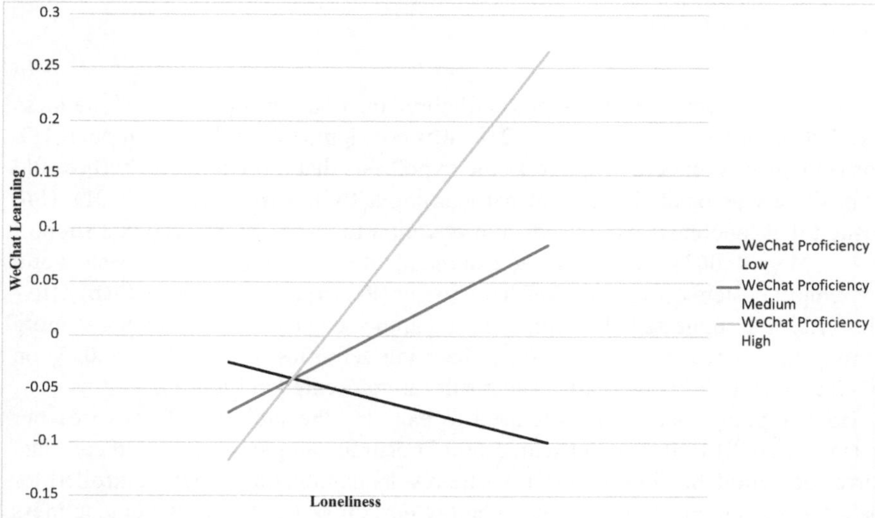

Figure 2.2 Interaction Effect.

Discussion

The present study examines how children's residential status affects Chinese midlife parents' loneliness, motivation to learn, and social media use. Prior U&G research has largely overlooked how the life challenges faced by this specific age group affect their social media use patterns (e.g., Tanis et al., 2017). The current study fills the remaining voids by examining how WeChat gaming and learning activities are driven by loneliness through different mechanisms, a characteristic of negative empty-nest feelings. Study findings demonstrate the applicability of U&G and SDT to Chinese middle-aged demographics. As loneliness is associated with negative health outcomes (review in Rico-Uribe et al., 2018), the present study has implications for both psychological consulting and online education industries.

Specifically, Chinese midlife empty-nest parents were lonelier than those with children at home. Although prior research suggests that Chinese empty-nest elders are not necessarily unhappier (Zhang, 2020; Zhang et al., 2021), the conclusion did not hold when it comes to loneliness among midlife parents. It seems that midlife parents need some time for role transitions when children leave home (e.g., Tanis et al., 2017). Moreover, prior research on empty nesters has mostly investigated negative emotions such as distress and depression (Mitchell & Lovegreen, 2009; Xu & Yang, 2023), while the current study calls for attention to loneliness among midlife empty-nesters.

Contrary to expectations, empty-nest mothers were not found to be lonelier than fathers. This finding again suggests that loneliness differs from other well-being variables (Mitchell & Lovegreen, 2009; Zhang, 2012). One of the reasons may be that women are kin keepers having closer connections with other family members than do men (Hagestad, 1986). They may thus receive more emotional compensation from interactions with other relatives. Additionally, the pervasive use of instant messenger may have become the equalizer that allows empty-nest parents to easily get in touch with their children. As suggested by Xu and Yang (2023), regular contact with children likely benefits the mental health of older empty nesters in China in the long run. Since social presence negatively mediates the impact of social media use on loneliness among older adults in China (Yang et al., 2021), WeChat interactive affordances—such as video calls, various stickers and emojis, tickling function, and so forth—could potentially help midlife empty-nesters (especially mothers) reduce loneliness.

Furthermore, the current research represents one of the first studies that established a negative link between midlife empty nest and WeChat learning behavior. This could be explained by the fact that Chinese lifelong learning features more social utilitarianism than individual interests and hobbies (Chen & Liu, 2019). As suggested by our study, empty-nest parents are older than non-empty nesters. They may hold senior positions in the workplace or even have retired, which according to SDT, partially explains their flagging motivation for WeChat learning.

In line with U&G, lonelier midlife parents played more WeChat mini-games than less-lonely parents. This echoes the finding by Chen and Leung (2016) that loneliness significantly increased time spent on mobile social games. As passing time has been cited as a major gratification of digital media use sought by lonely people (e.g., Fokkema & Knipscheer, 2007), when attention is focused on a moderately complex game, lonely people tend to perceive time to pass faster (Brown, 2008). The finding that gaming was favored by lonelier midlife parents than learning thus suggests that WeChat games might directly help said group regulate emotions by taking their minds off loneliness. However, learning provides a sense of meaning and boosts perceived competence, which can benefit midlife parents' psychological empowerment and facilitate successful aging in the long run (Chen et al., 2023). Therefore, longitudinal studies that involve more participants from diverse socio-economic backgrounds are warranted, as such variables influence learning behavior and life quality among older Chinese empty nesters (Zhang et al., 2021).

Although lonely and less-lonely parents did not differ in WeChat learning, interestingly, for the most proficient WeChat users, the more lonelier they were, the more learning they conducted on WeChat. This finding has implications, as engaging in learning activities seems to be an antidote against loneliness for technologically proficient midlife parents. This manifests how the one-stop service provided by WeChat fulfills Chinese users' different psychological needs. As lifelong learning improves older adults' psychological well-being (e.g., Narushima et al., 2016), it is important to improve elderly people's technology proficiency, so that they can engage in a broader range of activities to maintain cognitive abilities (Rowe & Khan, 1997) and mental health (Chen et al., 2023).

Finally, as expected, while motivation to learn improves engagement in learning activities on WeChat, loneliness reduces midlife parents' motivation to learn. To wit, motivation to learn completely mediates the relationship between loneliness and WeChat learning activities. This finding, alongside the moderating effect of WeChat proficiency, has complicated the relationship between loneliness and WeChat learning. One explanation could be that while loneliness reduces motivation to take on challenging learning tasks, higher WeChat proficiency enables users to conduct more incidental learning, which is more relaxing and fun. Future studies should consider how different types of learning interact with psychological factors—such as needs for autonomy and competence—and impact midlife parents' learning motivations and social media use.

Limitations and Future Research

There are a few limitations that could be addressed by future research. First, it was challenging to approach a non-student population. Due to different educational backgrounds, many respondents had never participated in an academic survey and were thus unable to fully understand the purpose of the current study or the meaning of certain questions (e.g., attention checks), leading to a higher dropout rate and more missing data. Snowball sampling might also cause nesting in data, which requires heed in later work. Second, the current sample does not have an ideally equivalent gender percentage, since women may be more likely to offer help when someone they know invites them to participate in the study (e.g., Ipsos, 2020). Empty-nest parents also failed to turn out as did non-empty nesters. Ideally more data should have been collected in a second wave. However, we decided not to resume data collection. This is because the U-turn in China's COVID policy changed people's residential status, with more college students returning home from school to live with their parents, who thus no longer remained in an empty nest. Additionally, fruit merging mini-games (e.g., Watermelon Merging, "合成大西瓜") on WeChat gained traction when we collected data, but soon lost momentum. Such historical events would have hurt the internal validity of the current study if we had collected another wave of data. Future research should aim to recruit more representative samples.

Third, a social desirability bias might exist in respondents' self-report of gaming and family loneliness. For example, negative connotations associated with online gaming could make individuals feel guilty of identifying themselves as gamers (Shaw, 2012). Such concerns might have led respondents to underestimate their gaming activities. Moreover, psychosocial well-being and social support from adult children are considered important dimensions of successful aging among elderly Chinese (Zhang et al., 2018), so people might be motivated to deflate their self-report family loneliness scores. Fourth, our study treated midlife empty-nest parents and non-empty nesters as homogenous, without further differentiating their family income, educational attainment, place of residence, marital status, and when their children will leave or have left home. These factors would affect how midlife parents allocate their attentional resources to a variety of activities and how well they have adapted or prepared to adapt to the leaving of their children; for instance, Tanis et al. (2017) observed that child-related social media use increased when the child was leaving the parental home while gradually decreasing over the subsequent two years, eventually to a level similar to that of six months before the child's leaving. Additionally, with watching short videos emerging as a common pastime for midlife empty-nest couples (Zhu, 2021), how short video applications (e.g., Douyin) are used to manage loneliness is also worth studying. Furthermore, in an increasingly urbanized, individualized, and Westernized society, marital relationship has gradually outweighed the parent–child relationship, as Zhang (2020) found that Chinese elders living without a spouse tend to be unhappier, while those living with or without children are equally happy; this finding calls for a potential re-definition of an empty nest as living without children and a spouse. Future researchers would benefit from including the above-mentioned variables in their analyses.

References

Alberti, F. B. (2019). *A biography of loneliness: The history of an emotion.* Oxford University Press.

Allen, M. S., Iliescu, D., & Greiff, S. (2022). Single item measures in psychological science: A call to action [Editorial]. *European Journal of Psychological Assessment, 38*(1), 1–5. https://doi.org/10.1027/1015-5759/a000699

Alleyne, R. (2010, March 16). *Middle age begins at 35 and ends at 58. The Telegraph.* www.telegraph.co.uk/news/health/news/7458147/Middle-age-begins-at-35-and-ends-at-58.html

Baumeister, R. F., DeWall, C. N., Ciarocco, N. J., & Twenge, J. M. (2005). Social exclusion impairs self-regulation. *Journal of Personality and Social Psychology, 88*(4), 589–604. https://doi.org/10.1037/0022-3514.88.4.589

Bek, H. (2017). Understanding the effect of loneliness on academic participation and success among international university students. *Journal of Education and Practice, 8*(14), 46–50.

Brown, S. W. (2008). Time and attention: Review of the literature. In S. Grondin (Ed.), *Psychology of time* (pp.111–138). Emerald Group.

Chen, C., & Leung, L. (2016). Are you addicted to Candy Crush Saga? An exploratory study linking psychological factors to mobile social game addiction. *Telematics and Informatics, 33*, 1155–1166. https://doi.org/10.1016/j.tele.2015.11.005

Chen, F. (2016, March 30). *The revolution of education in China—WeChat brings us back the mentorship education model.* LinkedIn. www.linkedin.com/pulse/revolution-education-china-wechat-becoming-biggest-digital-fang-chen/

Chen, S.-T., Wu, T.-Y., Atkin, D., Lee, P.-H., & Lai, Y.-C. *A greeting a day keeps the elderly Okay: Predicting the effects of sharing and generating the elderly's memes on successful aging.* [Manuscript submitted for publication, 2023]. Institute of Communication Studies, National Yang Ming Chiao Tung University.

Chen, Z., & Liu, Y. (2019). The different style of lifelong learning in China and the USA based on influencing motivations and factors. *International Journal of Educational Research, 95*, 13–25. https://doi.org/10.1016/j.ijer.2019.03.005

China's National Bureau of Statistics. (2012, April 27). 2011年我国农民工调查监测报告. www.stats.gov.cn/ztjc/ztfx/fxbg/201204/t20120427_16154.html

Chinese Social Science Net. (2015). *微信社会资本对集体行为影响报告*. Retrieved from www.cssn.cn/.

De Schutter, B., Brown, J., & Abeele, V. (2015). The domestication of digital games in the lives of older adults. *New Media & Society, 17*(7), 1170–1186. https://doi.org/10.1177/1461444814522945

Deci, E. L., & Ryan, R. M. (1984). *Intrinsic motivation and self-determination in human behavior*. Plenum Press.

Demetrovics, Z., Urbán, R., Nagygyörgy, K., Farkas, J., Zilahy, D., Mervó, B., Reindl, A., Ágoston, C., Kertész, A., & Harmath, E. (2011).Why do you play? The development of the motives for online gaming questionnaire (MOGQ). *Behavior Research Methods, 43*, 814–825. https://doi.org/10.3758/s13428-011-0091-y

Di'Tommaso, E., & Spinner, B. (1993). The development and initial validation of the Social and Emotional Loneliness Scale for Adults (SELSA). *Personality and Individual Differences, 14*(1), 127–134. https://doi.org/10.1016/0191-8869(93)90182-3

Dolan, R., Conduit, J., Fahy, J., & Goodman, S. (2016). Social media engagement behavior: A uses and gratifications perspective. *Journal of Strategic Marketing, 24*(3–4), 261–277. https://doi.org/10.1080/0965254X.2015.1095222

Duncan, S., & Edwards, R. (1999). *Lone mothers, paid work and gendered moral rationalities*. Macmillan.

Feng, X. (2009). 第一代独生子女父母的家庭结构:全国五大城市的调查分析. *Social Science Research* (Chinese), *4*, 104–110.

Finn, S., & Gorr, M. B. (1988). Social isolation and social support as correlates of television viewing motivations. *Communication Research, 15*(2), 135–158. https://doi.org/10.1177/009365088015002002

Fokkema, T., & Knipscheer, K. (2007). Escape loneliness by going digital: A quantitative and qualitative evaluation of a Dutch experiment in using ECT to overcome loneliness among older adults. *Aging & Mental Health, 11*(5), 496–504. https://doi.org/10.1080/13607860701366129

Goldscheider, F., & Goldscheider, C. (1999). *The changing transition to adulthood: Leaving and returning home.* Sage.

Gorges, J., Maehler, D. B., Koch, T., & Offerhaus, J. (2016). Who likes to learn new things: Measuring adult motivation to learn with PIAAC data from 21 countries. *Large-scale Assessments in Education, 4*(9), 1–22. https://doi.org/10.1186/s40536-016-0024-4

Hagestad, G. O. (1986). The family: Women and grandparents as kin-keepers. In A. Pifer & L. Bronte (Eds.), *Our Aging Society: Paradox and Promise* (pp. 141–160). W.W. Norton.

Harkins, E. B. (1978). Effects of empty nest transition on self-report of psychological and physical well-being. *Journal of Marriage and Family*, *40*(3), 549–556. https://doi.org/10.2307/350935

Hawkley, L. C., & Cacioppo, J. T. (2010). Loneliness matters: A theoretical and empirical review of consequences and mechanisms. *Annals of Behavioral Medicine, 40*(2), 218–227. https://doi.org/10.1007/s12160-010-9210-8

Hayes, A. F. (2017). *Introduction to mediation, moderation, and conditional process analysis* (2nd ed.). The Guilford Press.

Hu, L. T., & Bentler, P. M. (1999). Cutoff criteria for fit indexes in covariance structure analysis: Conventional criteria versus new alternatives. *Structural Equation Modeling, 6*(1), 1–55. http://dx.doi.org/10.1080/10705519909540118

Huang, H., & Zhang, X. (2017). The adoption and use of WeChat among middle-aged residents in urban China. *Chinese Journal of Communication, 10*(2), 134–156. https://doi.org/10.1080/17544750.2016.1211545

Ipsos (2020, April 3). Women more likely than men to have tried to help others amid COVID-19 outbreak. IPSOS. www.ipsos.com/en-uk/covid-19-coronavirus-outbreak-crisis-polls-community-support-help-women-men

Katz, E., Gurevitch, M., & Haas, H. (1973). On the use of mass media for important things. *American Sociological Review, 38*(2), 164–181. https://doi.org/10.2307/2094393

Kim, S. J., & Hancock, J. T. (2015). Optimistic bias and Facebook use: self-other discrepancies about potential risks and benefits of Facebook use. *Cyberpsychology, behavior and social networking, 18*(4), 214–220. https://doi.org/10.1089/cyber.2014.0656

Kline, R. B. (2015). *Principles and practice of structural equation modeling* (4th ed.). The Guilford Press.

Larmer, B. (2017, February 2). The parachute generation. *New York Times*. www.nytimes.com/2017/02/02/magazine/the-parachute-generation.html

Lerner, D. (1958). The Passing of traditional society: *Modernizing the Middle East*. Free Press.

Lev-on, A. (2017). Uses and gratifications: Evidence for various media. In P. Rössler, C. A. Hoffner, & L. van Zoonen (Eds.), *The International Encyclopedia of Media Effects* (pp. 1–9). John Wiley. https://doi.org/10.1002/9781118783764

Mitchell, B. A., & Lovegreen, L. D. (2009). The empty nest syndrome in midlife families. A multimethod exploration of parental gender differences and cultural dynamics. *Journal of Family Issues*, *30*(12), 1651–1670. https://doi.org/10.1177/0192513X09339020

Mwilambwe-Tshilobo, L, Ge, T., Chong, M., Ferguson, M. A., Misic, B., Burrow, A. L., Leahy, R. M., & Spreng, R. N. (2019). Loneliness and meaning in life are reflected in the intrinsic network architecture of the brain. *Social Cognitive and Affective Neuroscience, 14*(4), 423–433. https://doi.org/10.1093/scan/nsz021

Narushima, M., Liu, J., & Diestelkamp, N. (2016). Lifelong learning in active ageing discourse: Its conserving effect on well-being, health and vulnerability. *Ageing and Society, 38*(4), 651–675. https://doi.org/10.1017/S0144686X16001136

Olson, K. E., O'Brien, M. A., Rogers, W. A., & Charness, N. (2011). Diffusion of technology: Frequency of use for younger and older adults. *Ageing International, 36*(1), 123–145.

Pallavicini, F., Pepe, A., Mantovani, F. (2022). The effects of playing video games on stress, anxiety, depression, loneliness, and gaming disorder during the early stages of the COVID-19 pandemic: PRISMA systematic review. *Cyberpsychology, Behavior, and Social Networking, 25*(6), 334–354. https://doi.org/10.1089/cyber.2021.0252

Petik, K., Kezdy, A., & Kocsis, F. (2013). Learning projects and their background motivations. *European Journal of Medical Health, 8*(2), 187–211. https://doi.org/10.5708/EJMH.8.2013.2.2

Preacher, K. J., & Hayes, A. F. (2008). *Assessing mediation in communication research* (pp. 13–54). The Sage sourcebook of advanced data analysis methods for communication research.

Quinn, K. (2014). An ecological approach to privacy: "Doing" online privacy at midlife. *Journal of Broadcasting & Electronic Media, 58*(4), 562–580. https://doi.org/10.1080/08838151.2014.966357

Rico-Uribe, L. A., Caballero, F. F., Martín-María, N., Cabello, M., Ayuso-Mateos, J. L., & Miret, M. (2018). Association of loneliness with all-cause mortality: A meta-analysis. *PLoS ONE, 13*(1), e0190033. https://doi.org/10.1371/journal.pone.0190033

Rowe, J. W., & Kahn, R. L. (1997). Successful aging. *The Gerontologist, 37*(4), 433–440.

Ryan, R. M., & Deci, E. L. (2000). The darker and brighter sides of human existence: Basic psychological needs as a unifying concept. *Psychological Inquiry, 11*(4), 319–338. https://doi.org/10.1207/S15327965PLI1104_03

Schafer, M. H., & Shippee, T. P. (2010). Age identity in context: Stress and the subjective side of aging. *Social Psychology Quarterly, 73*(3), 245–264. https://doi.org/10.1177/0190272510379751

Schell, R., Hausknecht, S., Zhang, F., & Kaufman, D. (2016). Social benefits of playing Wii Bowling for older adults. *Games and Culture, 11*(1–2), 81–103. https://doi.org/10.1177/1555412015607313

Shaw, A. (2012). Do you identify as a gamer? Gender, race, sexuality, and gamer identity. *New Media & Society, 14*(1), 28–44.

Siegel, D. J. (2001). Toward an interpersonal neurobiology of the developing mind: Attachment relationships, "mindsight," and neural integration. *Infant Mental Health Journal, 22*(1–2), 67–94. https://doi.org/10.1002/1097-0355(200101/04)22:1<67::AID-IMHJ3>3.0.CO;2-G.

Statista. (2022). *Number of monthly active WeChat users from 2nd quarter 2011 to 1st quarter 2022 (in millions).* www.statista.com/statistics/255778/number-of-active-wechat-messenger-accounts/

Tanis, M., van der Louw, M., & Buijzen, M. (2017). From empty nest to social networking site: What happens in cyberspace when children are launched from the parental home? *Computers in Human Behavior, 68*, 56–63. https://doi.org/10.1016/j.chb.2016.11.005

Wang, G., Hu, M., Xiao, S., & Zhou, L. (2017). Loneliness and depression among rural empty-nest elderly adults in Liuyang, China: A cross-sectional study. *BMJ Open*, *7*(10), e016091. https://doi.org/10.1136/bmjopen-2017-016091

Wang, X. (2017). *新媒体环境下城乡中年群体媒介素养比较研究*. [Master's thesis, Zhejiang University of Technology]. CNKI.

Wang, Z., Tchernev, J. M., & Solloway, T. (2012). A dynamic longitudinal examination of social media use, needs, and gratifications among college students. *Computers in Human Behavior, 28*(5), 1829–1839. https://doi.org/10.1016/j.chb.2012.05.001

Wolfers, L. N., & Schneider, F. M. (2021). Using media for coping: A scoping review. *Communication Research, 48*(8), 1210–1234. https://doi.org/10.1177/0093650220939778

Xu, M., & Yang, W. (2023). Depressive symptoms among older empty nesters in China: The moderating effects of social contact and contact with one's children. *Aging & Mental Health, 27*(1), 54–69. https://doi.org/10.1080/13607863.2021.2019190

Yang, S., Huang, L., Zhang, Y., Zhang, P., & Zhao, Y. C. (2021). Unraveling the links between active and passive social media usage and seniors' loneliness: A field study in aging care communities. *Internet Research, 31*(6), 2167–2189.

Yang, Y., Crook, C., & O'Malley, C. (2014). Can a social networking site support afterschool group learning of Mandarin? *Learning, Media and Technology, 39*(3), 267–282. https://doi.org/10.1080/17439884.2013.839564

Zhang, C., Cai, Y., Xue, Y., Zheng, X., Yang, X., Lu, J., Hou, L., & Li, M. (2021). Exploring the influencing factors of quality of life among the empty nesters in Shanxi, China: A structural equation model. *Health Qual Life Outcomes, 19*, 156. https://doi.org/10.1186/s12955-021-01793-x

Zhang, S. (2012). The impact of college freshmen's leaving home on middle-aged empty-nest parents' psychological health. *Journal of Changchun Education Institute, 28*(11), 48–49.

Zhang, W. (2018). 微信小程序游戏的发展现状及其社交属性. *Journal of News Research (Chinese), 14*, 108.

Zhang, W., Liu, S., & Wu, B. (2018). Defining successful aging: Perceptions from elderly Chinese in Hawai'i. *Gerontology and Geriatric Medicine, 4*, 1–7. https://doi.org/10.1177/2333721418778182

Zhang, Y. (2020). Are empty-nest elders unhappy? Re-examining Chinese Empty-nest elders' subjective well-being considering social changes. *Frontiers in Psychology, 11*, 885. https://doi.org/10.3389/fpsyg.2020.00885

Zhu, Z. (2021). *中年空巢夫妻情感现状与调适研究*. [Master's thesis, Zhengzhou University]. CNKI.

3 Social Media Portrayal of Housewives and Gender Issues in Chinese Society

A Perspective of Digital Feminism Framework[1]

Ming Xie and Chin-Chung Chao

Introduction

In December 2021, a public divorce between singer Leehom Wang and his wife Jinglei Lee attracted people's attention on social media. Their divorce dispute dominated social media in Chinese' society for over a week. On December 15, 2021, Wang announced that he was divorcing Jinglei Lee on Sina Weibo, where Wang has over 67 million fans. The announcement received over 100,000 comments and 3.5 million likes on Sina Weibo. On December 17, Lee published a lengthy post on Weibo, where she has three million followers, and accused Wang of wrongdoings during their marriage. The post received 11.9 million likes within two days. In this post, Lee explained her sacrifice for their marriage, including giving up her own career and taking care of their three children. Lee's post raised the issue of how women often face unfair marriage life and financial income inequality, as well as Wang being an absent father (Zuo, 2021). Lee's words and posts were described as "蕾神之锤 (the hammer of Thunder Goddess)" because of her precise and powerful post and following response with strong writing skills.

Lee's post described the gender inequality faced by many housewives in Chinese society and presented a modern concept about the role of women in a patriarchal society. It also raised extensive public discussions on housewives and gender issues on various social media platforms. The present research focuses on public discussions on social media regarding this divorce drama and discusses how housewives and gender issues are portrayed on social media in Chinese society. Wang and Lee's divorce serves as an essential case study to understand and interrogate the current notions of feminist discourse and their interaction with digital technology in contemporary China. Based on gender schema theory (Starr & Zurbriggen, 2017), this research provides insights on the current feminist discourse as well as social reality regarding gender perception, women's role, and marriage on social media in Chinese society. It also helps to understand how social media in China have shaped feminist discourses and activisms.

In the following sections, we first discuss the theoretical framework: gender schema theory, the existing literature on feminist discourse in China, and digital feminism. After introducing the research design and research methods, we present the thematic analysis results of online discussions of Wang and Lee's divorce. The

DOI: 10.4324/9781003328896-4

chapter is concluded by summarizing the role of social media in promoting feminism in contemporary Chinese society.

Literature Review: Gender Schema Theory

Gender schema theory is a social-cognitive theory exploring the impact of gender on people's cognitive and categorical processing in daily lives (Starr & Zurbriggen, 2017). The theory proposes that individuals develop perceptions about gender schemas such as masculine and feminine and use these perceptions to categorize information, make decisions, and regulate behaviors. According to Bem (1981), gender is a cultural construction. In a society like the United States, gender polarizing is dominant in its discourse and social institutions. Therefore, children become gender schematic or gender polarizing, evaluate and process information based on the cultural definitions of gender appropriateness, and "reject any way of behaving that does not match their sex" (Bem, 1981, p. 126). Social institutions then reinforce the gender schema. Therefore, discussion of gender issues must be grounded in understanding cultural, social, and economic processes (Leung, 2003).

Gender schema theory focuses on how cultures construct gender and gender schemas. Also, it discusses how social institutions reinforce gender identities and roles. For example, mass media have mainly depicted women as housewives. Since Bem developed the theory in 1981, it has been widely used in various disciplines. Studies have explored how gender schema impacted career cognition and issues within the workplace (Chang & Hitchon, 2004; Chen et al., 2014). As Wood (2009) noted, "What gender means depends heavily on cultural values and practices; a culture's definitions of masculinity and femininity shape expectations about how individual men and women should communicate; and how individuals communicate establishes gender that, in turn, influence cultural views" (p. 20). Gender schema is constructed by cultural values and socio-demographic characteristics. It influences individuals' perceptions regarding individual's identity, as well as romantic relationships and marriage (Alani et al., 2016). Laosethakul and Leingpibul (2010) studied gender perception of the computing-related career/education fields and found that Chinese women faced challenges to be recognized and successful in a male-dominated society, although they have been expected to be as capable as men. Regarding housewives' role and domestic labor, it has been primarily viewed as unwaged labor. Housewives' domestic and affective labor is unpaid and subject to exploitation because it creates uncommodified products such as emotional care (Tan et al., 2020).

In China, gender schema is deeply rooted in Confucianism, with the notions such as "Men are born superior to women," "a woman without talent is a virtuous one" (Liu, 2006). The Communist Party of China adopted the one-child policy from the late 1970s until 2015. Some people believe that this policy might create more education and employment opportunities for girls and women. The reality is that most people still hold the gender binary expectations and stereotypes (Liu, 2006). As Liu (2006) argued, "gender-specific expectations for girls have been extended rather than fundamentally changed while those for boys remain largely the same as in the past." With social development, girls are expected to integrate both masculine and feminine characteristics.

Feminist Discourse in Chinese Society

In addition to gender schema theory, Hofstede (1984) also suggested that gender perceptions reflected cultural values. In both the United States and China, gender has been polarized as masculinity and femininity. The Chinese society has been historically a male-dominant society in which women's role is primarily in families to take care of husbands and children (Xiang et al., 2015). Women have been expected to be "virtuous wife and good mother" (Liu, 2014, p. 19). Also, Liu (2014) summarized the traditional Chinese female virtues such as female gentleness and female beauty and claimed that women with feminine characteristics and attributes of caregiver, reproducers, and nurturers were much more successful in the marriage market. With the social and economic change and globalization, the role and self-perception of women in China have been changed profoundly. On the one hand, Chinese women have been viewed as an important and equal part of the workforce. Chinese women believe that they are equal to men (Trauth et al., 2008). On the other hand, gender inequalities and discrimination still exist, and women have faced more challenges and difficulties for work–life balance and career development.

Feminism has been defined as "a movement to end sexism, sexist exploitation, and oppression" (Hooks, 2000, p. 1). Since the early 1990s, feminism has become a popular discourse in China. Chinese feminists started embracing the global concept of feminism to address gender issues in China. Concepts such as women's empowerment and women-centered sustainable development emerged in the mainstream discourse (Wang & Zhang, 2010). Scholars criticized the term "equality between men and women" promoted by the Communist Party of China because it simply claims that men and women can perform the same service for the patriarchal state (Wang & Zhang, 2010). Therefore, Chinese feminists adopted the term "gender equality," explicated gender-power relations, and called for attention to the limitations and constraints of women's liberation.

Feminism emphasizes women's growing awareness of the oppression and exploitation in the existing social structure and power relations. However, scholars also claimed that within a neoliberal paradigm, the claimed individual choice, empowerment, and voice have nothing related to structural injustice and economic redistribution (Wu & Dong, 2019). Neoliberal feminism advocates an independent gender relationship that women and men are complementary to each other (Rottenberg, 2018). Under the marketization, female sexuality and the public discourse of nurturing children through mothering have called married women to return home to fulfill their domestic duties. Women's familiar obligations have remained to be the main concept in Chinese society (Yin, 2021). Women are expected to be able to manage work-family conflict. Also, Peng (2021) argued that neoliberal feminism rationalized the complementary roles of women and men and promoted the feminized male ideal in an interdependent gender relationship.

Moreover, the popularity of social media and information technology may help reach what Peng proposed.

Social Media and Digital Feminism

With the prevalence of information technology and social media in daily life, scholars have studied how gender is performed in this domain. With the emergence of social media websites such as Facebook, Twitter, Sina Weibo, and WeChat, individuals have had opportunities to communicate widely and express gendered identities in the digital space.

Existing literature has mainly focused on the gender difference in online platforms. Strano (2008) studied gender differences in impression management and found that women engage in impression management more than men. The study found that women used their profile pictures more than men for the purpose of impression management. Samp et al. (2003) examined gender-swapping and gender monitoring at online platforms and found that gender schema was not associated with these practices. Rose et al. (2012) also studied Facebook users' profile pictures and found that the prominent traits for males' pictures were active, dominant, and independent, while females were attractive and dependent.

With the "Chinese feminist awakening," information technology provides channels and platforms for every user to make commentaries on celebrity cheating affairs and marriages (Kaiman, 2016; Wu & Dong, 2019). The Internet becomes the main domain to mediate conflicts that occur regarding romantic relationships and familiar issues.

Digital feminism is a concept focusing on the interaction between feminism and digital technology (Chang & Tian, 2021). Scholars discussed how digital technology created new feminist expression and feminist culture and mobilized public awareness of gender equality (Banet-Weiser & Portwood-Stacer, 2017; Wang & Driscoll, 2018). For example, scholars have explored how women around the world use social media to fight sexism, sexual harassment, and gender inequality (Jackson, 2018; Rentschler, 2014). Studies on various social media platforms such as Twitter, Tumblr, and YouTube described how digital technology provided opportunities for contemporary feminism's development (Clark, 2016; Rentschler, 2014).

With the proliferation of information technology, individuals have more possibilities to express their opinions regarding gender identity and gender roles as well as their social relationships. Lemons and Parzinger (2007) found that women in information technology professions were much less gender schematic than men. The question here is whether social media simply replicate cultural and mass media normative versions of gender roles and identity, or social media can help us understand gender roles more from a diverse and inclusive perspective.

Overall, studies have explored how people felt and perceived their own identities and duties in a family through cultural and feminist perspectives. This research sought to gain a deeper understanding of the interplay between the current feminist discourse, social media, and gender through gender schema theory (Bem,

1981). This research wanted to answer the specific question of how social media has portrayed housewives and reflected the gender schema changes in Chinese society.

Methodology

To answer the research question, this research adopted critical discourse analysis as a qualitative content analysis approach to analyze an online poll conducted by the website Yahoo. The critical discourse analysis approach puts the texts within a specific socio-cultural context and analyzes the texts at the textual, discursive, and socio-cultural levels (Fairclough, 2003; Khosravinik & Unger, 2016).

Right after Lee's post on Sina Weibo, Yahoo did a poll asking the question, "Do you think housewives should get paid?" Under the question, there was a blurb that stated,

> In Leehom Wang's divorce controversy, his ex-wife JingLei Lee published an article. In this article, Lee mentioned that modern women had the capacity to receive higher education while they could only play roles as wives and mothers in families. She said, "the salary of this job should include the opportunity cost that you have the capability, but you don't work."

Before the poll closed on December 20, 2021, there were 10,221 people who voted. Among these votes, 9,377 voted for "Yes, housewives make significant contributions to families," and 844 voted for "No, housework should not be paid." All the votes have comments.

We downloaded all the comments, deleted comments that only stated simple words such as "yes," "no," "of course," and "definitely." In total, 5,994 comments, including 5,385 comments from individuals who voted "yes" and 609 comments from those who voted "no," were analyzed for the present study. These comments might not be able to represent all Chinese people's views and opinions regarding the gender issues in China; they provide a window into a general discourse among the public. All the comments were in Mandarin and translated into English for data analysis. The two researchers are fluent in both Mandarin and English.

Data analysis adopted a theme analysis approach and focused on the comments' perception of themselves, housewives' role, family, and marriage. The two researchers first read the comments separately and repeatedly to identify the significant and repetitive statements. Then the two researchers discussed the significant statements and categorized them into groups, developing a structure of themes for data analysis. In the process, we allowed for emerging themes or concepts to arise. The final themes were constantly compared with the collected data and the literature to derive the discussions and conclusions of this research. The discursive analysis allowed the researchers to interpret the texts, connect the texts with the socio-cultural processes, and examine the contextual discourse production within wider society (Fairclough, 2003).

Findings

Three major themes emerged from the data: (1) traditional gender roles and stereotypes in contemporary China; (2) equality, empowerment, and social justice; (3) and feminist identity: struggles between a collective and individual identity. Intercoder reliability confirmed that these themes were consistent between the two researchers. Attitudes, identities, and culture are all keys to understanding the current discourse of feminism and gender role in Chinese society. Also, these themes are consistent with Bem's (1981) gender schema theory. The online comments reflected people's experiences, identities, and their own attitudes on the similarities and differences in gender roles.

Theme 1: Traditional Gender Roles and Stereotypes in Contemporary China

Traditional cultural expectations for women and gender role stereotypes emerged through ideas about marriage and family and reflected the cultural expectation of women's role in Chinese society. As mentioned in the literature, women have been viewed as the key to family, community, even state stability because women assume the main responsibility to take care of the husband, children, and the parents-in-law. One comment claimed that "housewives should have a salary because they are nurturing the next generation for the country." Another one stated, "housewives should get paid because they are the main force to stabilize our society." From a traditional perspective of Chinese culture, one of the women's core capacities is housetraining the husband and children. As one of the comments stated, "I am a housewife, and I take care of my husband, children, and my family very well. My husband gives me almost 50,000 New Taiwan dollars. This is what I deserve. And he also has the ability to make so much money. If you are a true man, you should learn from my husband." This statement presented a gender stereotype and a feminized male idea in Chinese society that men are breadwinners who should make enough money to support the family.

Although housewives were viewed as playing a key role in maintaining the stability of family, community, and society, domestic labor is gendered and valued less in society. Therefore, people who perform domestic labor are also viewed as less valuable. Traditional Chinese culture views men as dominant and women as subordinate. As one of the commenters who voted for "No" stated, "most men in Taiwan would say, we provide food and house for our wife. And the wife still wants salary? How could it be fair?" Housewives' main roles are taking care of children, husband, parents-in-law. A commenter who voted for "Yes" stated that nobody wanted to be a housewife because housewives were viewed by society as a parasite. A commenter echoed and mentioned, "Housewives' job is harder than anyone else. I would rather go to work than stay at home to be a housewife. People would look me down and say that I rely on my husband." Both the people who voted for "Yes" and "No" reflected a Chinese culture that views women as playing the main role in housework while the husbands are involved very little and do not care too much about the housework. Another comment also described the difficulties of being a housewife,

> Housewives take good care of the family, almost all year round, and are under great pressure. However, if the male protagonist pays the salary or the family, he often tramples on the housewife's dignity like charity. And the parents-in-law always treat the housewife as if she just relies on their son. It is better for girls to make money on their own.

In Chinese society, the involvement of the parent generation in a family is very common. A commenter stated, "In early days, the mother-in-law would force the wife to stay at home so that she could oppress the daughter-in-law. Some parents not only oppress the wife by themselves; they also make their son oppress his wife." A dichotomization of feminine and masculine work contextualizes these comments. A commenter noted,

> I used to be a housewife myself. When I buy a new dress, my mother-in-law would say that my husband [works] very hard to make money. My husband opened the windows so that dust was inevitable, but he said that I was idle at home and could not mop up the ground clean. The baby was born. For half a year, I was unable to sleep well because I was breastfeeding, and once I got a fever to 39 degrees and vomited. So I asked my husband to take care of the baby before midnight. He said that he needed to work the next day, and I could stay home and take a nap.

Nowadays, many women have had equal education opportunities and work capacity like men. However, housewives are viewed as material and property, even gold diggers, when they ask for money from their husbands. The negative attitude neglects their invisible contribution in maintaining the family's stability and taking care of seniors and children at home.

Theme 2: Equality, Empowerment, and Social Justice

The feminist discourse in China has emphasized the terms such as equality, empowerment, and social justice and defined Chinese women as autonomous and modern with the desire for self-expression (Peng, 2021). On the one hand, women in China are expected to fulfill the family obligations of taking care of husband and children; on the other hand, a feminist agenda has constantly proposed that women should be self-reliant and independent because it would be too risky for them to rely on their male partners. This kind of discourse reflects the deep-rooted gender inequality in China's socio-economic structure (Wallis & Shen, 2018).

The collected comments reflected a progressive narrative of gender issues in Chinese society. Some comments stated that contemporary Chinese women are no longer living in ancient times but in the twenty-first century. With sufficient education and career opportunities, women should pursue their own happiness and true love. For people who voted that housewives should get paid, one of the main justifications was respect and fairness. Some commenters mentioned that getting a salary is a way to respond to a human being's basic values. A comment noted that "the husband's family cannot just take for granted, even ignore the housewives' contribution."

The term reciprocity was mentioned many times in these comments. And getting payment for housewives' work is a way to ensure reciprocity. A commenter noted, "the contribution to a family should be equal between wife and husband. If it is not equal, there should be a payment." Another commenter echoed the statement, "how can a man work hard without his wife help him take care of the family and support him?"

However, most of the comments have not addressed anything related to the patriarchal socio-economic structure of Chinese society. There is only one comment that discussed the current family structure and socio-economic environment,

> The family structure of today has undergone significant changes. The labor market is no longer solely dependent on labor. The popularization of higher education has also created more job opportunities in the workplace. However, workplace equality for men and women is another topic. After the child is born, whether it is a housewife or a househusband, there should be a quantitative economic calculation regarding the family's contribution. Full-time housewives and househusbands should have the so-called opportunity cost. And not only the financial resources are contributions. But human nature always sees that money is more than time spent. Therefore, in the current state of society, each opportunity cost should be calculated reasonably to calculate the remuneration due to full-time housewives and househusbands and to establish this remuneration as a social standard so that the work of full-time mothers or fathers can be valued.

In addition to gender issues, another term that emerged from the comments is class. Wang's family was viewed as superior to Lee's. Therefore, Wang's class is higher than Lee's. The rationalization of gender hierarchy adheres to a patriarchal objectification of women in the current Chinese society.

Another controversy in these comments is about whether women should give up their career opportunities for their families. People who voted for "yes" stated that "housewives give up the opportunities for their own ability development and financial independence to take care of family. So, the other person who is making money should compensate the housewife." In contrast, people who voted "no" commented,

> Some people get married just because they don't want to work. They want to find a long-term meal ticket. So, it depends on the situation. Suppose they have a job and are forced to resign because of the need of the family or children; they might need to get the salary. But for people whose job is not stable, they should not get paid.

Another commenter expressed a similar idea,

> Modern women have the ability to receive higher education, but they can only play the roles of wives and mothers in the family. This is entirely a personal choice. I don't understand why they don't go out to work if they have the ability. Professional women sacrifice the time they spend with their children when they grow up. Financial independence is the same as self-realization. Professional women have to face the problem of mother-in-law and daughter-in-law as they do housework after getting off work. Why can housewives enjoy both? There is a profession called babysitter, and you can give your baby to a babysitter. Take it; if you go out to work, no one will force you, unless you communicate badly with your partner, don't tell me it is not safe to have the babysitter take care of your child. Professional women are taking this kind of worry and risk!!! In short, you have chosen to be a housewife. You should bear your own choice, or you just need to go out and find a job!

The divorce between Wang and Lee attracted so much public attention because of the topic on housewives. Although some people believe that being a housewife is a personal choice, the current public discourse has promoted the idea that staying in the workforce and keeping the competitiveness on the job market is the safest choice for women. Compared to many Western countries, women's employment rate in Chinese society is much lower, especially when compared to women in different age groups. Statistics show that in 2018, women's employment rate in Hong Kong was 55.1- percent, while men's rate was 68.3 percent. And around 35 percent of women could not join the workforce because of the need to take care of their families. Taiwan is a similar situation. In 2020, women's employment rate was 51.41 percent, while men's was 67.24 percent. In addition, the employment rate for women at the age 55–59 in Taiwan was only 35, while it was 65 percent in Japan, 72 percent in Germany, and 82 percent in Sweden (本尼, 2021). The longer time women spend time on housework, the less opportunity they will have in the job market, which truly hinders the development of gender equality in Chinese society.

Theme 3: Feminist Identity: Struggles between Collective and Individual Identities

Traditional Confucianism constructs women by defining their relations with men as mothers, sisters, and wives. Therefore, women are secondary within the collectives of the family and community (Leung, 2003). Family responsibilities impose a double burden on women. With the purpose to maintain a harmonious family and state stability, women are expected and required to take responsibility and to be benevolent and caring for others.

On the contrary, Chinese women have also been encouraged to seek individual liberation and emancipation through their own discourse, self-determination, and self-awareness raising. There is a clear shift in women's gender identities, combining women's domestic role, reproductive role, nurturing role, and work role. Among the commenters, women's self-reliance has been mentioned many times. The commenters asserted that women should live for themselves and should not sacrifice their life and happiness for their husband, partner, or family. Women should be outspoken, independent. One commenter stated, "The most important part for housewives is the recognition of self-value. Many people think that only working outside makes contributions. Actually, dealing with all the big and small stuff in the family requires significant endurance."

The emphasis on women's role in the family also creates another debate regarding whether housewives should get paid, which is the power relation between the wife and husband and the associated impact caused by payment. People who think housewives should get a salary stated that the salary is a symbol of recognition, respect, and fairness. However, people who think housewives should not be paid claimed that if she gets a salary from the husband, then the husband becomes the boss or employer, which would make the power relation between them more imbalanced. As one of the commenters argued, "if the housewife should get a salary, should we set up KPI [Key Performance Indicator]? So the wife should do housework based on the husband's standard. Just like in the workplace, the standard is not set by yourself."

> From a neoliberal feminist perspective, women should take the self-responsibility for their own well-being. A commenter claimed that the payment for housewives' work should be for their efforts and contribution but not a kind of alms. Under this situation, economic gain is much more important than anything else. Some people believe that housewives should not get paid: women make a choice because of their love for the family. However, another comment stated, "the love for a family cannot be exchanged for anything." Also, not everything in the family can be quantified and assessed by financial measures.
>
> A comment connected the discussion with the feminist discourse and claimed,
>
> [i]t is a conflict with the statement of "no gender stereotype and no gender discrimination" that is always shouted by some people. I would vote for yes if we were in a feudal society now. But we are in modern society. We can have a nanny and technology to help with the housework. And if both wife and husband can make money, they do not need to rely on anybody. If my wife truly wants to stay at home, I would spend money on a nanny, washing machine, and so on. But I would hope my wife could learn something about investment. So I can give my money for her to manage and invest.

Gender relations in China have gone through significant evolvement due to social, economic, and political changes. Chinese women's new identity involves aspects of their role in the patriarchal family and social relations. The discussion on whether housewives should be paid reflects the various opinions held by different people. One of the core arguments is over whether housework is a kind of job. People who think the housewife is doing harder and more difficult work than most other jobs claimed that housewives should definitely receive salary as a recognition of their contribution to the family and society. On the contrary, some people believe that housework should be the responsibility of both wife and husband, and there is no way to measure the contribution by payment.

Discussion and Conclusion

Overall, the prevalence of social media has made the representations of gender inequality in daily lives more explicit. The online discussions raised the public's awareness regarding how the traditional gender schema has been normalized, and become a dominant discourse in social life. Social media also provide spaces for women to share their experiences and opinions and to inspire and support each other beyond the location and environmental limitations. As Sadowski (2016) claimed, the current feminist activism became a digital-material entanglement with the impact of information technology.

In romantic relationships and marriage, not everything can be measured by financial methods. According to some of the comments, women generally feel the most uncomfortable because their contribution to the family is taken for granted by their husband, family, and society. Their housework is not considered to have the same work pressure as making money to support a family. As some of the commenters who voted for "No" stated, housewives spend a lot of time at home, watching movies and shopping, and it is an easy job. During the COVID-19 pandemic, many people are working from home. A comment stated, "the husband work at his company, and he does not need to do anything when he gets home. But the wife is working from home, without time off." According to the statistics of UN Women, the time that women spent on parenting was six hours longer than men. During the COVID-19 pandemic, this gap increased to 7.7 hours, and women averagely spent 31.5 hours on parenting, which is almost another full-time job (本尼, 2021). Among the collected comments, there were more than 9,000 who voted "Yes," which means they believe housewives should get paid, and no more than 800 votes supported that housewives should not get paid. However, among the 9,000 comments, we can see there are still many people believe women's main job should be taking care of family and being a good housewife. Although gender equality has been the main term in public discourse, housewives' work still receives much less recognition due to the gender schema and the traditional perception of gender roles in Chinese society.

Social media have empowered women to express their voices and challenge the status quo, prejudices, and power imbalances. In the process of seeking social justice, social media and digital platforms have been an effective approach for women to advocate feminist ideas among the general public. Jinglei Lee took a non-confrontational approach to express her feelings and requests. This approach is also praised by many people. The online discussion also reflects that women's progress is still hindered in the male-centered society of China. The traditional

Chinese culture has formed the segregation of gender roles that women should take care of and be devoted to family (Hofstede, 1984). Women are always more likely to sacrifice their careers. The present research revealed the deeply rooted gender stereotypes and their overall influences in Chinese society regarding gender identity and gender roles. As previous studies have identified, the contemporary feminist culture in China did not explicitly confront patriarchal norms (Chang et al., 2018; Chang & Tian, 2021). Since the sphere of housework is not solely occupied by women, and some men have taken their domestic duties as "househusbands," it is interesting to continue such a research journey with the question: Are the househusbands viewed/expected similarly to housewives in media, especially social media?

Note

1 The article is revised based on the article published in *China Media Research* Volume 18, Issue 1, 2022, pp. 102–114. Reproduced with permission from *China Media Research.*

References

Alani, A., Clark-Taylor, A., Rogeshefsky, A., & Cerulli, C. (2016). Exploring collegiate perceptions of feminism through Bem's gender schema theory. *New York Journal of Student Affairs*, *16*(1), 20.

Banet-Weiser, S., & Portwood-Stacer, L. (2017). The traffic in feminism: An introduction to the commentary and criticism on popular feminism. *Feminist Media Studies*, *17*(5), 884–888.

Bem, S. L. (1981). Gender schema theory: A cognitive account of sex typing. *Psychological Review*, *88*(4), 354–364. https://doi.org/10.1037/0033-295X.88.4.354

Chang, C., & Hitchon, J. C. B. (2004). When does gender count? Further insights into gender schematic processing of female candidates' political Advertisements. *Sex Roles*, *51*(3), 197–208. https://doi.org/10.1023/B:SERS.0000037763.47986.c2

Chang, J., & Tian, H. (2021). Girl power in boy love: Yaoi, online female counterculture, and digital feminism in China. *Feminist Media Studies*, *21*(4), 604–620. https://doi.org/10.1080/14680777.2020.1803942

Chang, J., Ren, H., & Yang, Q. (2018). A virtual gender asylum? The social media profile picture, young Chinese women's self-empowerment, and the emergence of a Chinese digital feminism. *International Journal of Cultural Studies*, *21*(3), 325–340.

Chen, Y.-C., Lee, C.-S., Yu, T.-H., & Shen, J.-Y. (2014). Effects of gender role and family support on work adjustment among male flight attendants in Taiwan. *Social Behavior and Personality: An International Journal*, *42*(3), 453–464. https://doi.org/10.2224/sbp.2014.42.3.453

Clark, R. (2016). "Hope in a hashtag": The discursive activism of #WhyIStayed. *Feminist Media Studies*, *16*(5), 788–804.

Fairclough, N. (2003). *Analysing Discourse: Textual analysis for social research*. Routledge.

Hofstede, G. (1984). *Culture's Consequences: International differences in work-related values* (Abridged edition). SAGE.

Hooks, B. (2000). *Feminism is for everybody*. Pluto Press. www.plutobooks.com/9780745317335/feminism-is-for-everybody

Jackson, S. (2018). Young feminists, feminism and digital media. *Feminism & Psychology, 28*(1), 32–49. https://doi.org/10.1177/0959353517716952

Kaiman, J. (2016, August 19). *A Hollywood-style celebrity scandal is dominating the Internet in China. Los Angeles Times*. www.latimes.com/world/asia/la-fg-china-celebrity-scandal-20160819-snap-story.html

Khosravinik, M., & Unger, J. W. (2016). Critical discourse studies and social media: Power, resistance, and critique in changing media ecologies. In R. Wodak & Mi. Meyer (Eds.), *Methods for critical discourse analysis* (pp. 206–233). Sage.

Laosethakul, K., & Leingpibul, T. (2010). Why females do not choose computing? A lesson learned from China. *Multicultural Education & Technology Journal, 4*(3), 173–187. https://doi.org/10.1108/17504971011075174

Lemons, M. A., & Parzinger, M. (2007). Gender schemas: A cognitive explanation of discrimination of women in technology. *Journal of Business and Psychology, 22*(1), 91–98. https://doi.org/10.1007/s10869-007-9050-0

Leung, A. S. M. (2003). Feminism in transition: Chinese culture, ideology and the development of the women's movement in China. *Asia Pacific Journal of Management, 20*, 359–374.

Liu, F. (2006). Boys as only-children and girls as only-children—Parental gendered expectations of the only-child in the nuclear Chinese family in present-day China. *Gender and Education, 18*(5), 491–505. https://doi.org/10.1080/09540250600881626

Liu, F. (2014). From degendering to (re)gendering the self: Chinese youth negotiating modern womanhood. *Gender and Education, 26*(1), 18–34. https://doi.org/10.1080/09540253.2013.860432

Peng, A. Y. (2021). Neoliberal feminism, gender relations, and a feminized male ideal in China: A critical discourse analysis of Mimeng's WeChat posts. *Feminist Media Studies, 21*(1), 115–131. https://doi.org/10.1080/14680777.2019.1653350

Rentschler, C. A. (2014). Rape culture and the feminist politics of social media. *Girlhood Studies, 7*(1), 65–82. https://doi.org/10.3167/ghs.2014.070106

Rose, J., Mackey-Kallis, S., Shyles, L., Barry, K., Biagini, D., Hart, C., & Jack, L. (2012). Face it: The impact of gender on social media images. *Communication Quarterly, 60*(5), 588–607. https://doi.org/10.1080/01463373.2012.725005

Rottenberg, C. (2018). *The rise of neoliberal feminism*. Oxford University Press.

Sadowski, H. (2016). From #aufschrei to hatr.org: Digital–material entanglements in the context of German digital feminist activisms. *Feminist Media Studies, 16*(1), 55–69. https://doi.org/10.1080/14680777.2015.1093090

Samp, J. A., Wittenberg, E. M., & Gillett, D. L. (2003). Presenting and monitoring a gender-defined self on the Internet. *Communication Research Reports, 20*(1), 1–12. https://doi.org/10.1080/08824090309388794

Starr, C. R., & Zurbriggen, E. L. (2017). Sandra Bem's gender schema theory after 34 years: A review of its reach and impact. *Sex Roles, 76*(9), 566–578. https://doi.org/10.1007/s11199-016-0591-4

Strano, M. (2008). User descriptions and interpretations of self-presentation through Facebook profile images. *Journal of Psychosocial Research on Cyberspace, 2*(2), Article 5.

Tan, C., Wang, J., Wangzhu, S., Xu, J., & Zhu, C. (2020). The real digital housewives of China's Kuaishou video-sharing and live-streaming app. *Media, Culture & Society, 42*(7–8), 1243–1259.

Trauth, E. M., Quesenberry, J. L., & Huang, H. (2008). A multicultural analysis of factors influencing career choice for women in the information technology workforce. *Journal of Global Information Management*, *16*(4), 1–23. https://doi.org/10.4018/jgim.2008100101

Wallis, C., & Shen, Y. (2018). The SK-II #changedestiny campaign and the limits of commodity activism for women's equality in neo/non-liberal China. *Critical Studies in Media Communication*, *35*(4), 376–389. https://doi.org/10.1080/15295036.2018.1475745

Wang, B., & Driscoll, C. (2018). Chinese feminists on social media: Articulating different voices, building strategic alliances. *Continuum*, *33*, 1–15. https://doi.org/10.1080/10304312.2018.1532492

Wang, Z., & Zhang, Y. (2010). Global concepts, local practices: Chinese feminism since the fourth UN conference on women. *Feminist Studies*, *36*(1), 40–70.

Wood, J. T. (2009). *Gendered Lives: Communication, gender, and culture* (8th ed.). Wadsworth.

Wu, A. X., & Dong, Y. (2019). What is made-in-China feminism(s)? Gender discontent and class friction in post-socialist China. *Critical Asian Studies*, *51*(4), 471–492. https://doi.org/10.1080/14672715.2019.1656538

Xiang, Y., Ribbens, B., Fu, L., & Cheng, W. (2015). Variation in career and workplace attitudes by generation, gender, and culture differences in career perceptions in the United States and China. *Employee Relations; Bradford*, *37*(1), 66–82. http://dx.doi.org.leo.lib.unomaha.edu/10.1108/ER-01-2014-0005

Yin, S. (2021). Re-articulating feminisms: A theoretical critique of feminist struggles and discourse in historical and contemporary China. *Cultural Studies*, online first, 1–25. https://doi.org/10.1080/09502386.2021.1944242

Zuo, M. (2021, December 22). Estranged wife of Wang Leehom emerges as 'goddess' for Chinese writing skills. *South China Morning Post*. www.scmp.com/news/people-culture/china-personalities/article/3160706/wang-leehom-saga-lee-jinglei-emerges

本尼 (2021). 王力宏李靚蕾離婚事件牽出的華人全職「家庭主婦」處境討論. BBC News, www.bbc.com/zhongwen/trad/chinese-news-59766596

4 Facebook-Based Mental Health Discourse in Bangladesh

Self-Disclosure, Social Support, Consultation

Simu Dey and Josh Averbeck

Introduction

Stressful life circumstances are considered the ultimate cause of psychological distress, and self-disclosure, nonetheless, could buffer the harmful consequences of stress on mental health. The study by Zhang (2017) indicates that people can self-disclose on social network sites (SNSs), despite being ambiguous to what degree they benefit from accomplishing that. However, self-disclosure can be beneficial for an individual on social media platforms such as Facebook, Instagram, and Twitter. Recently, social media sites have emerged as increasingly adopted platforms wherein health information-seeking and sharing practices are apparent. Distinct from online fora, these social systems are more holistic because millions of people use them to post about the mundane goings of their lives. In addition, most social media platforms have a permanent, unique identifier tied to each user's account, creating a rich environment for examining various aspects of self-disclosure, social support, and disinhibition in shaping discussions around health topics (De Choudhury & De, 2014).

This chapter focuses on at-risk populations in Bangladesh to understand the impact and effect of mental health support through social media. According to a 2019 survey by *The Daily Prothom Alo*, nearly 17 percent of adults in Bangladesh suffer from mental health issues, and 92 percent of them do not seek professional help (*The Daily Prothom Alo*, 2021). Facebook is one of the largest social network sites whose user base includes people of all ages and backgrounds. Several investigations explored the content and role of Facebook groups pertinent to different chronic diseases such as diabetes, breast cancer, and colorectal cancer, and risk factors such as tobacco use (Bender et al., 2011; De la Torre-Díez & Diaz-Pernas, 2012; Greene et al., 2011; Hefler et al., 2012; Zhang et al., 2013). However, there is no study on mental health and Facebook groups and how people can get support from these groups. This present study analyzes the content of Facebook group messages for mental health support and the relationship between the message type and its effect on the feelings of support experienced by message recipients and impact.

This study aimed to understand how different Facebook groups can help people to understand their mental stress. Mental health-focused social media groups and pages allow people to disclose their inherent feelings and agony. This study investigates

DOI: 10.4324/9781003328896-5

whether people seeking mental health support on these social media platforms do not want to reveal their information from their real Facebook accounts and how people get more emotional support on a post. Because religion remains important in Bangladesh, religious support in private groups will be examined and compared to the feedback in public groups. Furthermore, the relationship between analytical thinking, clout, authenticity, and emotional tone for Facebook posts by mental health support seekers will be considered, as they can be essential even though there may be differences in support types on public pages versus private groups. This study focused on those who own electronic devices and use Facebook to access mental health services and information and their implications for mental health equity.

Mental Health in Bangladesh

With social networking and the advancement of individualized medical care, patients share their health information through social networking sites such as Facebook (Househ, 2011). To understand how Facebook enables Bangladeshi people to increase their access to information and improve their well-being, one must explore the factors influencing whether people have access to and use digital mental health information and services.

A cross-sectional study of health systems in Bangladesh found a lack of resources and infrastructure to provide optimal care for mental health patients (Nuri et al., 2018). Furthermore, Alam et al. (2020) found that only 220 psychiatrists and 50 trained clinical psychologists serve Bangladesh, with a population of roughly 163 million. This lack of resources would be troubling for any nation. Bangladesh has endured recent stressors that have increased mental health concerns. The dense population, natural disasters, and an increase in Myanmar refugees have led to cases of mental health disorders. The cross-sectional study by Riley et al. (2017) investigated trauma history, daily environmental stressors, and mental health effects for 148 Rohingya adults living in Kutupalong and Nayapara refugee camps in Bangladesh. Their study also found high mental health crises such as posttraumatic stress disorder (PTSD), depression, somatic complaints, and related practical impairment. The results by Riley et al. (2017) indicated that the participants approved of local idioms of distress, including somatic complaints and crises correlated with spirit possession. Coupled with the lack of resources, the people of Bangladesh have turned to the Internet for informal mental health treatment (Alam et al., 2020).

Several studies about Facebook have examined its provision of health information (Ahmed et al., 2020; Alam et al., 2020; Nuri et al., 2018). However, none of these has researched psychological treatment and suggestions via Facebook groups and fan pages regarding mental health problems in Bangladesh, although mental illness is still stigmatized in society. Because people still struggle to differentiate mental health and mental disorder (Islam & Biswas, 2015). The current study might fill the gap by measuring the real pictures of mental health guidance for people using Facebook. This research might benefit individuals and society of Bangladesh to mitigate the mental health problem. Specifically, individuals share and read about mental health information on Facebook concerning social support for their health behaviors. Furthermore, those people want to get support from Facebook groups and pages.

This study focuses on the behavior of mental health support seekers in public and private virtual spaces and the emotional and religious support they receive. The study examines whether people tend to remain anonymous in public virtual spaces and whether negative posts are associated with more emotional support. Additionally, religious support is also more prevalent in private groups, while people react more negatively to feedback on public pages. The study also explores the relationship between analytical thinking, clout, authenticity, and emotional tone in Facebook posts by mental health support seekers and the variations in support types on pages versus groups. The overarching question guides the study of how people use Facebook pages and groups to get social support, and what kind of support are they getting.

Social Support Theory

This study employed the Social Support (SS) theory framework that explains how people support each other within a social support network. The base point of the Social Support (SS) theory developed by Don Drennon-Gala and Francis Cullen is that every person living in society receives various kinds of support from the social support network (SSN) (Eskandari, & Ghahramanloo, 2020). According to SS theory, individuals derive psychological and emotional support from their social relationships. According to this theory, individuals who feel supported by their social networks have improved mental health and well-being and are less likely to experience stress and anxiety. This SS theory helps us to understand how and what types of social support are provided by Facebook pages and groups to Bangladeshi users, because SS theory suggests that support can come in different forms, including emotional, informational, and tangible support. Individuals must have access to multiple sources of support to achieve optimal mental health outcomes. A strong social networking site (SNSs) such as Facebook provides people with the opportunity of receiving social support (Hwang et al., 2010; Winzelberg et al., 2003; Wong, 2016). SNS-based SS is more effective than face-to-face because SNSs provide some unique aspects of supportive communication, such as anonymity, non-judgmental interaction, and amenities (Hwang et al., 2010). In Bangladesh, some popular mental health-related groups and pages have developed a strong SSN and provide social support to their followers.

This study used the core logic of Linguistic Inquiry and Word Count (LIWC) originated from decades of scientific research indicating that people's language can provide deep insights into their psychological states, including their sentiments, feeling styles, and social problems. Sometimes, these insights are relatively prominent and specific. For instance, assume someone uses many words like *delighted*, *enthusiastic*, and *satisfied*. In that case, they probably feel happy, and we can use this information to estimate their current emotional state reliably. Frequently, however, the relationships between verbal behavior and psychology are much, much less noticeable. People who are more convinced and higher in social situations tend to use "you" words at reasonably high rates and "me" at moderately low rates. Here, too, decades of empirical analysis—significant research utilizing LIWC as a scientific tool—provide us with technological ways of understanding, describing, and quantifying psychological, social, and behavioral phenomena (LIWC website).

There is a huge distinction between a group and a page on Facebook. A *page* is a public profile that permits organizations and brands to present their administrations and exercises. A group is a local area-based component that assembles individuals with similar interests to talk about themes and offer their sentiments. A previous study on self-disclosure on social media found that most posts were anonymous, and 90 percent of the total post on self-disclosure (De Choudhury & De, 2014). Moreover, the results show that 16.7 percent of Facebook users disclose their mental health information anonymously, whereas 83.3 percent of users seek mental health information from Facebook groups and pages none anonymously.

The social stigma will limit the public disclosure of mental health concerns. That is why people feel more comfortable remaining anonymous in public virtual spaces than in private. Virtual spaces that others can easily access are considered public spaces, meaning there is no expectation of privacy. When information is shared, control over it is lost, and privacy depends on the recipient's confidentiality (Petronio, 2010). Online social spaces can be private or public, and social norms dictate how information is shared. Participants feel free to share sensitive information in private virtual spaces, while in public spaces, they should be careful and share only what they want to be widely known. Tverdek (2008) argues that information can become public by being revealed intentionally or unintentionally.

Social Media Based Social Support in Bangladesh

Self-disclosing refers to expressing one's well-being, suffering, and feelings to others. On Facebook-based platforms, mental health information seekers use self-disclosure to express their feelings and suffering stresses (De Choudhury & De, 2014). People searching for mental health information from Facebook-based groups use it to narrate their sufferings in their posts. People who seek mental health information on Facebook-based mental health-related pages use it to express his/her sufferings in a comment box. However, the significant social stigma associated with mental health in Bangladesh has an adverse effect on seeking treatment (Islam & Biswas, 2015). Thus, our first hypothesis states:

H1: Mental health support seekers are more likely to remain anonymous in public virtual spaces than in private virtual spaces.

There is evidence that clout, authenticity, emotional tone, and analytical thinking are associated with mental health support seekers (Pennebaker, 1993). Individuals tend to use language that reflects their ability to influence others, honesty, emotions, and logic (Pennebaker et al., 2003). What is not yet clear is how these language behaviors are portrayed in social support messages. When a group member self-discloses information in a support group, the other members try to support the vulnerable person. Authenticity could be related to emotional tone, but the relationship among the other variables needs to be explored. Hence, this study is guided by our first research question:

RQ1: What is the relationship between analytical thinking, clout, authenticity, and emotional tone for Facebook posts by mental health support seekers?

When individuals post about their mental health using a more negative tone, other users will be more likely to respond with more emotional support. The negative emotional tone signals to other users that support is needed (Choudhury et al., 2013). Without this signal, users are less likely to offer social support. Thus, the second hypothesis is derived:

H2: Posts with a negative tone receive more emotional support.

As the SS theory explains, emotional support is offered on Facebook on both private groups and public pages. What is not clear is whether the types of support may differ on pages versus groups. Private groups may allow individuals to be more emotionally supportive because of their privacy. Public pages may feel more instructive and foster different kinds of support messages. Hence, the second research question is asked:

RQ2: Are there differences in support types on public pages versus private groups?

Bangladesh is considered a highly religious country (Alam et al., 2020). Thus, religious support is likely to be offered. However, religious displays are still considered a private matter (Alam et al., 2020). Therefore, religious support messages are more likely to appear in private groups than on public pages. Hence, the third hypothesis states:

H3: More religious support will be offered in private groups than on public pages.

Negative feedback is more likely to be shared in response to support messages on public pages than in private groups (Bazarova, 2012). Entry to private groups may limit access to those who wish to offer more negative responses (Alam et al., 2020). Public pages also tend to reach a wider audience, including those who would not typically offer support to individuals requesting it. So, our fourth hypothesis is:

H4: There will be more negative feedback on public pages than on private groups.

Methodological Approach

Research Sample

This research uses two public pages and a private group on Facebook. These were chosen because of their popularity and focus on mental health in Bangladesh. The name of the group is "Psychology." It is a private Facebook group for mental health issues, created four years ago and with 253,000 members as of April 10, 2021. Six administrators moderate the group. By answering some basic questions, anyone can join this private group. People who are group members can share their thoughts and experiences. "LifeSpring" is a Facebook page that discusses psychiatry and many other issues. This page had 409,925 followers as of April 10, 2021. The content of the page frequently reaches 30,000 to 35,000 viewers. LifeSpring is the page run by the society "LifeSpring." The society provides video content to its followers on its Facebook page. It often arranges a live video session named "Ask Your Psychiatrist," in which they discuss mental health issues, and people communicate with each other through commentary. Another Facebook page is "Dr. Shusama Reza." It is a fan page of Dr. Shusama Reza. She is a public figure, having 629,000 followers as of April 10, 2021. She often shares mental health-related content on her page, which receives thousands of likes and views and hundreds of comments.

These Facebook groups and pages have a large dataset of thousands of users' posts, comments, and other content. For this research, five posts and ten video comments of group members and page-followers, which are related to self-disclosure, social support, and consultation, were collected by taking screenshots. The contents that are not textual, such as photos, photovoice (descriptions of photographs to share experiences), videos, and live video sessions, were collected by direct downloading. Also, one hundred comments on these posts and videos were collected as research samples because, with a hundred comments, posts were considered as the most engaged with members.

Measures

Anonymity. Facebook users can post anonymously or be identified. Users' identification information was either provided or not. Among the observed Facebook accounts, 30 were considered, with 25 providing identifiable information, while the remaining five opted for anonymity.

Examining through the lens of anonymity as enabled by throwaway accounts on Facebook, we find that a small but significant fraction of Facebook users employ the feature as a cover for more personal and open discussions around their experiences of mental illness. In fact, despite the negative or caustic nature of content shared by anonymous Facebook users, online disinhibition of this nature garners more emotional and instrumental feedback through comments (De Choudhury & De, 2014).

Pronoun Use. The presence of first and third-person pronouns was counted using the software of psycholinguistic lexicon *Linguistic Inquiry and Word Count* (LIWC) (De Choudhury, 2014). Table 4.1 provides the descriptive for first-person pronouns and social words (the standard LIWC variable for talk of others).

LIWC Composite Variables. Analytical thinking, clout, authenticity, and emotional tone are summary variables calculated by LIWC. Analytical thinking is indicative of formal, logical thinking patterns. Clout reflects confidence and social status. First-person plural (we) and second person (you) are indicators of higher clout. Authenticity reflects word choices associated with honesty and vulnerability. The

Table 4.1 Linguistic Inquiry and Word Count Analysis Results

	Sum	*M*
First Person Pronoun	464	15.47
Social Words	307.1	10.24
Positive Emotion	76.6	2.55
Negative Emotion	136.9	4.56
Cognitive Process	351.6	11.72

Source: Table created by authors during analysis.

emotional tone reflects positive and negative emotional word usage. Higher scores are indicative of a positive emotional tone. (See Table 4.1).

Another linguistic measurement scale was used for measuring the dependent variable, s*ocial support*. Prior research on health and psychology identified four categories of social support, such as prescriptive, informational, instrumental, and emotional (De Chowdhury & De, 2014; George et al., 1989; Turner et al., 1983).

In Bangladesh, most people are religious (Alam et al., 2020). They often try to narrate their happiness and obligations in the light of religion. So, sometimes Bangladeshi mental health information seekers attain suggestions or support from a religious approach; this kind of support is considered religious or social support in this research. De Choudhury (2014) used a model that employs unigrams and bigrams present in posts, videos, and comments to automatically infer clusters of social support. For assessing different types of social support, we counted how many times a mental health information seeker receives comments that have the following unigram and bigram (see Table 4.2).

Primary Data

A total of fifteen samples were selected from these aimed Facebook groups and pages. Among them, five samples are sorted from the group "Psychology," and the rest were taken from two different pages, "LifeSpring" and "Dr. Shusama Reza." In groups, individuals wrote about their mental health concerns in a post, and others tried to support them in a comment box. On pages, doctors hold live video sessions, followers write their problems in a comment box, and doctors try to support them in live video. All the posts, videos, and comments were in Bengali; we translated them into English before analyzing them.

The Group: "Psychology" is dedicated to mental health issues. Members often write their problems and seek mental health support. The administrators of the group, as well as other people, are seen to support them in the comment box. Table 4.3 provides an overview of the topics, reactions, and comments for supporting the support seekers.

Mental Health-Related Facebook Pages. Qualified and skilled doctors run both pages, "LifeSpring" and "Dr. Shusama Reza." They provide mental health-related

Table 4.2 Examples of Support Coding

Emotional Support
Good luck! I am afraid, I do not have any advice. *Good luck!* I can guess your situation; it is really terrible. *Good luck!* Please, talk honestly to your doctor. *Good luck!* I can understand your suffering, I also suffer from it. You will come round soon.
Informational Support
Therapy, cognitive behavioral therapy, is very helpful in this situation. *Therapy*, with a therapist you can get very customized understanding. *Treatment,* you should go to … psychiatrist. *Treatment,* I know a skilled psychiatrist.
Instrumental Support
Feel free to knock me if you ever need. *Feel free* to look up what you have been prescribed. *…feel free*. I am here. *…feel free*. Come to my inbox.
Prescriptive Support
Medication, prevents seizures, lessens withdrawal. *Medication*, take this medicine and follow this instruction. *Medication*, mindful meditation will improve your mental situation. You can do YOGA.
Religious Support
God bless. You should pray to God. *God bless.* God will help you. *God bless.* Never be disappointed, God always remains with us.

Source: Table created by authors during analysis.

Table 4.3 Descriptive Data for the Group "Psychology"

Topic	*Dates*	*Reactions*	*Comments*
Schizophrenia	1 March, 2021	265	278
Depression	3 March, 2021	255	80
Adele syndrome	4 March, 2021	110	116
Thanatophobia	9 March, 2021	170	80
Inferiority complex	13 March, 2021	75	47

Source: Table created by authors during analysis.

Table 4.4 Descriptive Data for Videos

Headline	*Presenter*	*Duration*	*Date*	*Comments*	*Shares*	*Views*
Selfcare	Dr. Shusama Reza	28:51:00	1 December, 2020	2.8k	1.1k	133,000
Pillars of Life: Overcoming Depression; Episode: 18.	Dr. Sayedul Ashraf	19:46	25 December, 2020	120	318	28,000
Mindfulness Meditation	Dr. Shusama Reza	7:30	2 December, 2020	527	882	141,000
Am I the Cause of My Sorrows?	Dr. Munmun Jahan	24:53:00	11 March, 2021	129	384	45,000
Ask Your Psychiatrist: Insomnia	Dr. Rubaiyat Ferdush Dr. Shafiul Alam	34:43:00	9 March, 2021	106	38	10,000
Ask Your Psychiatrist	Dr. Shafiul Alam Dr. Golam Mostofa	41:26:00	9 February, 2021	107	13	5,500
Ask Your Psychiatrist	Dr. Shafiul Alam Dr. Rubaiyat Ferdush	31:06:00	4 February, 2021	68	13	6,300
Ask Your Psychiatrist	Dr. Rubaiyat Ferdush Dr. Golam Mostofa	40:58:00	28 January, 2021	63	16	12,000
Ask Your Psychiatrist	Dr. Shafiul Alam Dr. Kamrul Islam	53:19:00	5 September, 2020	107	14	11,000
What is Schizophrenia?	Dr. Golam Mostofa	3:56	18 January, 2021	64	356	62,000

Source: Table created by authors during analysis.

information on their pages and other health issues. These two pages provide video content to their followers on their Facebook pages. They often arrange live video sessions, titled "Ask Your Psychiatrist," where they discuss mental health issues, and people communicate with them through commentary. Five self-disclosing posts from the group "Psychology" and 25 self-disclosing comments from the pages "LifeSpring" and "Dr. Shusama Reza" are used to analyze this research. Table 4.4 provides descriptive data about the videos used in this study.

Results

Self-disclosing and Language

The first hypothesis predicted that mental health support seekers are more likely to remain anonymous in public virtual spaces than in private virtual spaces. A chi-square test for independence with $\alpha = .05$ was used to assess whether the types of self-disclosures on Facebook accounts were related to the medium of self-disclosing. The chi-square test was statistically significant, $\chi2$ (1, N =30) = 17.328, p <.0001, with a Phi (φ) coefficient of .76, indicating a large relationship. People who sought mental health information in groups were more likely to self-disclose from anonymous accounts than the people who used to seek mental health information on public pages. Thus, H1 was supported.

RQ1 asked about the relationships between analytical thinking, clout, authenticity, and emotional tone for Facebook posts by mental health support seekers. Tests were conducted to assess the size and direction of the liner relationship between the scores of analytical thinking, clout, authenticity, and emotional tone, bivariate Pearson's product-moment correlations (r). The bivariate correlation between two variables, clout and authenticity, was negative and strong, r (30) = −.70, $p < .001$. It indicated that the more authentic and honest language one used, the less clout or status one exhibited. There were no other significant correlations among the variables (see Table 4.5).

Emotional Support and Private vs. Public Spaces

Hypothesis 2 predicted that posts with a negative tone are associated with more emotional support. To assess the size and direction of the linear relationship between the scores of emotional tones in self-disclosure and in attainment of emotional support, a bivariate Pearson's product-moment correlation coefficient (r) was calculated. The bivariate correlation between these two variables was positive and strong, r (30) = .55, p<.002. Prior to calculating r, the assumptions of normality, linearity, and homoscedasticity were assessed and found to be supported. Specifically, a visual inspection of the normal Q-Q and de-trended Q-Q plots for each variable confirmed that both were normally distributed. Similarly, visual inspection of the scatterplot of attainment scores of emotional supports against scores of emotional tones confirmed that the relationship between these variables was linear and heteroscedastic. Thus, H2 was supported.

Table 4.5 Relationships among Composite LIWC Variables

	Authenticity	*Analytic Thinking*	*Clout*
Analytical Thinking	.42		
Clout	−.70**	.07	
Emotional Tone	.02	.08	−0.19

Source: Table created by authors during analysis.

Note: ** indicates $p < .001$.

RQ2 asked if there are differences in support types on public pages versus private groups. A series of independent sample t-tests were conducted. The independent sample t-test to compare the attainment of emotional support found that neither Shapiro-Wilk statistic was significant, indicating that the assumption of normality was not violated. Levene's test was also non-significant; thus, an equal variance can be assumed for both groups. The t-test was statistically significant, with the mean of attainment of emotional support from the group (M =14.6, SD = 12.42) significantly higher (mean difference 10.4, 95% CI [2.57, 18.22]) than the pages (M = 4.20, SD =6.7), t (28) =2.724, p<.001, two-tailed, Hedges's gs = .021. The common language (CL) effect size indicates that, the chance that for a randomly selected pair of individuals, the attainment score of emotional support from a group is higher than the attainment score from pages is 50 percent.

The independent sample t-test to compare the attainment of informational support also found that neither Shapiro-Wilk statistic was significant, indicating that the assumption of normality was not violated. Levene's test was also non-significant; thus, an equal variance can be assumed for both groups. The t-test was statistically significant, with the mean IAT score of males (M =8.8, SD =8.198) significantly higher (mean difference 7.28, 95% CI [4.034, 10.526]) than the females (M = 1.52, SD =1), t (28) =4.6, p<.001, two-tailed, Hedges's gs = 2.19. The common language (CL) effect size indicates that the chance that for a randomly selected pair of individuals, the IAT score of a male is higher than the score of a female is 81 percent.

Similarly, the independent sample t-test compares the attainment of instructive support. Neither Shapiro-Wilk statistic was significant, indicating that the assumption of normality was not violated. Levene's test was also non-significant; thus, an equal variance can be assumed for both groups. The t-test was statistically significant, with the mean attainment of instructive support score from the group (M =5, SD =1.581) was significantly higher (mean difference 3.480, 95% CI [2.509, 4.451]) than the pages (M = 1.52, SD =.823), t (28) =7.33, p<.05, two-tailed, Hedges's gs = 3.548. The common language (CL) effect size indicates a 97 percent likelihood that, when comparing a randomly selected pair of individuals, the instructional support score obtained from the group will be higher than the attainment score obtained from pages.

The independent sample t-test to compare the attainment of prescriptive support also found that neither Shapiro-Wilk statistic was significant, indicating that the assumption of normality was not violated. Levene's test was also non-significant;

thus, an equal variance can be assumed for both groups. The t-test was statistically significant, with the mean attainment of prescriptive support score from the group (*M*=8.8, *SD*=5.404) was significantly higher (mean difference 5.64, 95% CI [2.788, 8.492]) than the pages (*M* = 3.16, *SD* =2.135), *t* (28) =4.051, *p*<.001, two-tailed, Hedges's gs = 1.93. The common language (CL) effect size indicates the chance that for a randomly selected pair of individuals, the attainment of prescriptive support score from the group is higher than the attainment score from pages 83 percent.

Religious Support

Hypothesis 3 predicted there would be more religious support offered in private groups than on a public page. Facebook-based mental health-related platforms users are seen to provide religious and social support to their followers or members. During 1.9 visits that a Bangladeshi mental health information seeker gets religious support (Bhugra, 2004). An independent sample t-test was used to compare the attainment of religious support as a part of social support of the mental health information seekers in Facebook-based groups (*n* =5) and pages (*n* =25). Neither Shapiro-Wilk statistic was significant, indicating that the assumption of normality was not violated. Levene's test was also non-significant; thus, an equal variance can be assumed for both groups. The t-test was statistically significant, with the mean attainment of religious and social support score from the group (M=11.4, SD=6.348) was significantly higher (mean difference 11.4, 95% CI [8.992, 13.80]) than the pages (M = .00, SD =.00), t (28) =9.70, p<.001, two-tailed, Hedges's gs = 4.62. The common language (CL) effect size indicates that the chance for a randomly selected pair of individuals, the attainment of religious and social support score from the group is higher than the attainment score from pages (96%). Thus, Hypothesis 3 was supported.

Negative Feedback on Public Pages

Hypothesis 4 predicted there would be more negative feedback on public pages than on private groups. People who seek mental health information support are often seen to be victims of negative reactions such as negative comments, emojis, and so forth. Bangladeshi mental health information seekers faced a negative reaction on average 1.27 times on a public website (Sultana et al., 2021). An independent sample t-test was used to compare the attainment of negative reactions of the mental health information seekers in Facebook-based groups (*n* =5) and pages (*n* =25). Neither Shapiro-Wilk statistic was significant, indicating that the assumption of normality was not violated. Levene's test was also non-significant; thus, an equal variance can be assumed for both groups. The t-test was statistically significant, with the mean of attainment of negative reaction score from the group (*M* =7.4, *SD* =5.505) significantly higher (mean difference 7.36, 95% CI [5.264, 9.456]) than the pages (*M* = .04, *SD* =.20), *t* (28) =7.19, *p*<.001, two-tailed, Hedges's gs = 3.42. The common language (CL) effect size indicates that the chance for a randomly selected pair of individuals to attain a negative reaction score from the group is higher than the attainment score from pages (90%). In the

Facebook groups where doctors do not participate, their people get more negative comments. But with Facebook pages led by doctors, they always try to provide positive feedback for mental support. Thus, H4 was supported.

Discussion

This study examines how mental health support seekers are likelier to remain anonymous in public virtual spaces than in private virtual spaces and how posts with a negative tone are associated with more emotional support. This study further explores how more religious support is offered in private groups than on a public page and how people react to more negative feedback on public pages than in private groups. Finally, this study investigates the relationships between analytical thinking, clout, authenticity, and emotional tone for Facebook posts by mental health support seekers and what are the differences in support types on pages versus groups.

Not every person can manage the cost of a specialist, and not every person is happy with transparently discussing their medical problem. The vast majority seek counsel anonymously, or even to help somebody they know who is experiencing psychological sickness (Wahl, 1999). The primary motivation behind this investigation was to examine how individuals get various kinds of help from Facebook groups regarding mental health issues. However, this chapter explored the language that patients use online to convey their mental health needs and seek support from specialists and individuals who discuss mental health on a social network (Facebook). The investigation has utilized examples from more than 15 Facebook gatherings. Five examples of arranged gatherings were from a brain research group. Behavioral brain research considers understanding the biological basis of learning, memory, behavior, perception, and consciousness as the "ultimate challenge." In this gathering, however, individuals with psychological wellness issues seek help through messages, remarks, and posts. The makers of the gathering assist individuals with their issues through different methods.

Bangladesh is one of the major nations where individuals look for psychological wellness through Facebook groups and pages (De Choudhury & De, 2014). This study found that about 16.7 percent of people use anonymous social media accounts to get mental health information. In addition, this current study also revealed that an average of 15.4 percent of Bangladeshi mental health information seekers utilize first-person pronouns in recognizing their own issues to make their posts more interactive with their audiences, perhaps to attain suggestion and social support. Conversely, excessive use of first-person pronouns such as I, me, myself for high egocentrism (Pennebaker, & King, 1999). A highly egocentric focus is known as the psychological attribution of mental illness (Chung & Pennebaker, 2007).

Almost 50 percent of these individuals utilize psychological words that depict how they feel and what makes their circumstances harder, pushing them to the edge of their psychological prosperity. Other words have been especially identified with family companions, friends, and family. Bangladeshi mental health information seekers use a wide range of cognitive biases as well as negative emotions in narrating their mental stress. These observations are according to psychology literature. Dysfunctional attitudes, negative emotions, and depressive attributions manifest cognitive biases are the characteristics of mental illness of the health and social issues they face (Eaves & Rush, 1984).

It is not particularly simple to shout out about one's own issues, anonymously or not, and numerous individuals have maintained their reality while unveiling their issues through composition. The results of this study depicted that practically 85 percent of them attempted to lighten validity while the rest showed a more noteworthy passionate tone and helpfully introduced themselves. Patients acquire huge help through others' sympathy, empathy, and genuine concern. Advertising terms acknowledge their context and emotions and provide feedback by offering assistance and appropriate perspective guidance. This enthusiastic help is conceived out of genuine, valid associations with others. The findings of this study also showed that Bangladeshi mental health information seekers receive five types of social support: emotional support is average (5.97), informational (2.7), instructional (2.10), prescriptive (3.83), and religious (1.9). Most Bangladeshis are religious, so they suggest to other people, who are struggling with mental challenges, to practice religion. However, this suggestion is better only in Facebook groups, not the pages on which skilled doctors operate. Our significant findings show that the pages operated by professionals differ from groups with more non-professionals.

Limitations and Future Studies

We recognize that there are a few constraints in our study. This work was exclusively based on the comments collected through screenshots and videos from Facebook groups and pages. No constant members were deciding their emotions and boundaries; there were no meetings directed at hearing the thoughts of psychologists. An additional investigation should be possible through studies and meetings for focus groups or interviews. Also, this investigation was confined to just Facebook use in Bangladesh instead of utilizing a more extensive viewpoint, such as the utilization of other online media platforms and from different nations. Language plays a part in the interpretation, and there may be an opportunity for confusion because all the data collected from Facebook pages and groups were in Bengali.

Conclusion

This study investigated how mental health support seekers are likely to remain anonymous in public virtual spaces than in private virtual spaces. The results found that people prefer to use an anonymous account on Facebook to share their problems of mental illness. Moreover, people who use posts with a negative tone receive more emotional support from Facebook pages and groups. This study further explored people seeking mental support on Facebook groups and pages; they are offered more religious support in a private group than a public page. Nonetheless, people react to more negative feedback on public pages than in private groups. Finally, this study found significant relationships between analytical thinking, clout, authenticity, and emotional tone for Facebook posts by mental health support seekers.

References

Ahmed, T., Rizvi, S., Rasheed, S., Iqbal, M., Bhuiya, A., Standing, H., Bloom, G., & Waldman, L. (2020). Digital health and inequalities in access to health services in Bangladesh: Mixed methods study. *JMIR mHealth and uHealth, 8*(7), 2–10. https://doi.org/10.2196/16473

Al Mamun M., Ibrahim H. M., & Turin T.C. (2015). Social media in communicating health information: An analysis of Facebook groups related to hypertension. *Preventing Chronic Disease: Public Health Research, 12*(11). http://dx.doi.org/10.5888/pcd12.140265external icon

Alam, F., Hossain, R., Ahmed, H. U., Alam, M. T., Sarkar, M., & Halbriech, U. (2020). Stressors and mental health in Bangladesh: Current situation and future hopes. *BJPsych International. 18*(4). https://doi.org/10.1192/bji.2020.57

Bazarova, N. N. (2012). Public intimacy: Disclosure interpretation and social judgments on Facebook. *Journal of Communication, 62*(5), 815–832.

Bender, J. L., Jimenez-Marroquin, M. C., & Jadad, A. R. (2011). Seeking support on Facebook: A content analysis of breast cancer groups. *Journal of Medical Internet Research, 13*(1). https://doi.org/10.2196/jmir.1560

Bhugra, D. (2004). Migration and mental health. *Acta Psychiatrica Scandinavica, 109*(4), 243–258.

Chung, C., & Pennebaker, J. (2007). The Psychological Functions of Function Words. In K. Fiedler (Ed.), *Social communication* (pp. 343–359). Psychology Press. https://psycnet.apa.org/record/2007-01308-012

Cohen, R. A., & Adams, P. F. (2011). Use of the internet for health information: The United States, 2009. *NCHS Data Brief* (66), 1–8. https://pubmed.ncbi.nlm.nih.gov/22142942/

Cohen, S., & Wills, T. A. (1985). Stress, social support, and the buffering hypothesis. *Psychological Bulletin, 98*(2), 310–357. https://doi.org/10.1037/0033-2909.98.2.310

Corrigan P. (2004). How stigma interferes with mental health care. *The American Psychologist, 59*(7), 614–625. https://doi.org/10.1037/0003-066X.59.7.614

De Choudhury, M., & De, S. (2014, May). Mental health discourse on Reddit: Self-disclosure, social support, and anonymity. At the *Eighth International Association for the Advancement of Artificial Intelligence (AAAI) Conference on Weblogs and Social Media*, in Ann Arbor, MI.

De Choudhury, M., Gamon, M., Counts, S., & Horvitz, E. (2013). Predicting depression via social media. *Proceedings of the International AAAI Conference on Web and Social Media, 7*(1), 128–137. https://ojs.aaai.org/index.php/ICWSM/article/view/14432

De la Torre-Díez, I., Díaz-Pernas, F. J., & Antón-Rodríguez, M. (2012). A content analysis of chronic diseases social groups on Facebook and Twitter. *Telemedicine Journal and e-Health: The Official Journal of the American Telemedicine Association, 18*(6), 404–408. https://doi.org/10.1089/tmj.2011.0227

Eaves, G., & Rush, A. J. (1984). Cognitive patterns in symptomatic and remitted unipolar major depression. *Journal of Abnormal Psychology, 93*(1), 31–40. https://doi.org/10.1037//0021-843x.93.1.31

Eskandari, H., & Baratzadeh Ghahramanloo, N. (2020). Investigating the mediating role of social support in the relationship between addiction to social network, media literacy and emotional intelligence. *Journal of Cyberspace Studies, 4*(2), 129–151.

Eysenbach G. (2007). From intermediation to disintermediation and apomediation: New models for consumers to access and assess the credibility of health information in the age of Web2.0. *Studies in Health Technology and Informatics*, *129*(1), 162–166. https://pubmed.ncbi.nlm.nih.gov/17911699/

Fowler G. A. (2012, October 4). Facebook: One billion and counting. *The Wall Street Journal.* www.wsj.com/articles/SB10000872396390443635404578036164027386112

George, L. K., Blazer, D. G., Hughes, D. C., & Fowler, N. (1989). Social support and the outcome of major depression. *The British Journal of Psychiatry: The Journal of Mental Science*, *154(4)*, 478–485. https://doi.org/10.1192/bjp.154.4.478

Graham, S. (1991). A review of attribution theory in achievement contexts. *Educational Psychology Review*, *3*(1), 5–39. https://doi.org/10.1007/BF01323661

Greene, J. A., Choudhry, N. K., Kilabuk, E., & Shrank, W. H. (2011). Online social networking by patients with diabetes: A qualitative evaluation of communication with Facebook. *Journal of General Internal Medicine*, *26*(3), 287–292. https://doi.org/10.1007/s11606-010-1526-3

Harmon, J. (1958). *The Psychology of Interpersonal Relations*. By Fritz Heider. New York: John Wiley, *37*(3), 272–273, https://doi.org/10.2307/2572978

Hefler, M., Freeman, B., & Chapman, S. (2012). Tobacco control advocacy in the age of social media: using Facebook, Twitter and change. *Tobacco Control*, *22*(3), 210–214. https://doi.org/10.1136/tobaccocontrol-2012-050721

Househ M. (2011). Sharing sensitive personal health information through Facebook: The unintended consequences. *Studies in Health Technology and Informatics*, *169*, 616–620. https://pubmed.ncbi.nlm.nih.gov/21893822/

Hwang, K. O., Ottenbacher, A. J., Green, A. P., Cannon-Diehl, M. R., Richardson, O., Bernstam, E. V., & Thomas, E. J. (2010). Social support in an Internet weight loss community. *International Journal of Medical Informatics*, *79*(1), 5–13. https://doi.org/10.1016/j.ijmedinf.2009.10.003

Islam, A., & Biswas, T. (2015). Mental health and the health system in Bangladesh: Situation analysis of a neglected domain. *American Journal of Psychiatry and Neuroscience*, *3*, 57–62. https://doi.org/10.11648/j.ajpn.20150304.11

Johnson, G. J., & Ambrose, P. J. (2006). Neo-tribes: The power and potential of online communities in health care. *Communications of the ACM*, *49*(1), 107–113. https://cacm.acm.org/magazines/2006/1/6042-neo-tribes/fulltext

Kelley, H. H., & Michela, J. L. (1980). Attribution theory and research. *Annual Review of Psychology*, *31*(1), 457–501. www.annualreviews.org/doi/abs/10.1146/annurev.ps.31.020180.002325

LIWC website www.liwc.app/help/howitworks

Logan, A. G. (2014). Community hypertension programs in the age of mobile technology and social media. *American Journal of Hypertension*, *27*(8), 1033–1035. https://doi.org/10.1093/ajh/hpu125

Lupton, D. (2014). The commodification of patient opinion: The digital patient experience economy in the age of big data. *Sociology of Health & Illness*, *36*(6), 856–869. https://doi.org/10.1111/1467-9566.12109

Menefee, H. K., Thompson, M. J., Guterbock, T. M., Williams, I. C., & Valdez, R. S. (2016). Mechanisms of communicating health information through Facebook: Implications for consumer health information technology design. *Journal of Medical Internet Research*, *18*(8), 218. https://doi.org/10.2196/jmir.5949

Naslund, J. A., Aschbrenner, K. A., Marsch, L. A., & Bartels, S. J. (2016). The future of mental health care: Peer-to-peer support and social media. *Epidemiology and Psychiatric Sciences*, *25*(2), 113–122. https://doi.org/10.1017/S2045796015001067

Nuri, N. N., Sarker, M., Ahmed, H. U., Hossain, M. D., Beiersmann, C., & Jahn, A. (2018). Pathways to care of patients with mental health problems in Bangladesh. *International Journal of Mental Health Systems, 12(1),* 1–12. https://doi.org/10.1186/s13033-018-0218-y

Pennebaker, J. W. (1993). Putting stress into words: Health, linguistic, and therapeutic implications. *Behaviour Research and Therapy*, *31*(6), 539–548. https://doi.org/10.1016/0005-7967(93)90105-4

Pennebaker, J. W., & King, L. A. (1999). Linguistic styles: Language use as an individual difference. *Journal of Personality and Social Psychology*, *77*(6), 1296–1312. https://doi.org/10.1037/0022-3514.77.6.1296

Pennebaker, J. W., Mehl, M. R., & Niederhoffer, K. G. (2003). Psychological aspects of natural language use: Our words, our selves. *Annual Review of Psychology*, *54*(1), 547–577.

Petronio, S. (2010). Communication privacy management theory: What do we know about family privacy regulation? *Journal of Family Theory & Review*, *2*(3), 175–196.

Reddy, S. K., Mazhar, S., & Lencucha, R. (2018). The financial sustainability of the World Health Organization and the political economy of global health governance: A review of funding proposals. *Globalization and Health, 14*(1), 1–11. https://doi.org/10.1186/s12992-018-0436-8

Riley, A., Varner, A., Ventevogel, P., Taimur Hasan, M. M., & Welton-Mitchell, C. (2017). Daily stressors, trauma exposure, and mental health among stateless Rohingya refugees in Bangladesh. *Transcultural Psychiatry*, *54*(3), 304331.

Sayeed, A., Hassan, M. N., Rahman, M. H., El Hayek, S., Al Banna, M. H., Mallick, T., Hasan, A., Meem, A. E., & Kundu, S. (2020). Facebook addiction associated with internet activity, depression, and behavioral factors among university students in Bangladesh: A cross-sectional study. *Children and Youth Services Review, 118*(C), 1–6. https://doi.org/10.1016/j.childyouth.2020.105424

Sperlich, P. (1976). Attribution: Perceiving the causes of behavior. *American Political Science Review, 70*(2), 617–618. https://doi.org/10.2307/1959677

Sultana, M. S., Khan, A. H., Hossain, S., Islam, T., Hasan, M. T., Ahmed, H. U., ... & Khan, J. A. (2021). The association between financial hardship and mental health difficulties among adult wage earners during the COVID-19 pandemic in Bangladesh: Findings from a cross-sectional analysis. *Frontiers in Psychiatry*, *12*, 635884.

The Daily Prothom Alo. (2021, January 8). *Facebook partners with Bangladeshi orgs on mental heaith.* https://en.prothomalo.com/bangladesh/facebook-partners-with-bangladeshi-orgs-on-mental-health

The Daily Star. (2017, February 4), *LifeSpring to integrate mental, physical health care in Bangladesh.* www.thedailystar.net/lifestyle/lifespring-integrate-mental-physical-health-care-bangladesh-1355986

Turner, R. J., Frankel, B. G., & Levin, D. M. (1983). Social support: Conceptualization, measurement, and implications for mental health. *Research in Community & Mental Health, 3*, 67–111.

Tverdek, E. (2008). What Makes Information" Public"? *Public Affairs Quarterly*, *22*(1), 63–77.

Wahl, O. F. (1999). Mental health consumers' experience of stigma. *Schizophrenia Bulletin*, *25*(3), 467–478.

Waldman, L., Ahmed, T., Scott, N., Akter, S., Standing, H., & Rasheed, S. (2018). We have the Internet in our hands: Bangladeshi college students' use of ICTs for health information. *Globalization and Health*, *14*(1), 31. https://doi.org/10.1186/s12992-018-0349-6

Winzelberg, A. J., Classen, C., Alpers, G. W., Roberts, H., Koopman, C., Adams, R. E., & Taylor, C. B. (2003). Evaluation of an internet support group for women with primary breast cancer. *Cancer: Interdisciplinary International Journal of the American Cancer Society*, *97*(5), 1164–1173. https://doi.org/10.1002/cncr.11174

Wong, T. K. & Ma, W. W. K. (2016 June). Exploring the relationship between online social support and individual online subjective well-being among young adults. *Presented at 66th ICA Annual Conference*, at Fukuoka, Japan.

Zhang, R. (2017). The stress-buffering effect of self-disclosure on Facebook: An examination of stressful life events, social support, and mental health among college students. *Computers in Human Behavior*, *75*(2017), 527–537. https://doi.org/10.1016/j.chb.2017.05.043

Zhang, Y., He, D., & Sang, Y. (2013). Facebook as a platform for health information and communication: A case study of a diabetes group. *Journal of Medical Systems*, *37*(3), 1–12. https://doi.org/10.1007/s10916-013-9942-7

Part II

Mobile Communication in the Organizational Sphere

5 Mobile Communication as Disaster Response Infrastructure for Cross-Sector Coproduction

A Field Study of the Mobile Apps in China Flood Seasons[1]

Minshuai Ding

Introduction

In late May 2020, severe rainstorms hit many parts of China, resulting in the worst flooding since 1998 (Cheng, 2020). Despite this, flood-related deaths and missing persons in 2020 were at an all-time low, much lower than in 1998 (Ministry of Emergency Management of China, 2021). The increased usage of smartphone-based mobile applications (apps) for flood response is considered to be one of the factors that may have contributed to this improvement. During the deadly 2021 flood season in China, mobile apps remained prominent, with people using social media, particularly smartphone apps, to band together and support flood victims (BBC, 2021).

This technical phenomenon has emerged as a significant factor in disaster response. Mobile apps such as news, navigation, and social media have become increasingly important during disaster situations like bushfires, storms, and floods (Ding, 2022; American Red Cross, 2012; Cheong & Cheong, 2011; Latonero & Shklovski, 2011; Power et al., 2014). However, the functionalities of mobile apps used for disaster response in China have not been well-studied. While mobile apps were utilized in various aspects of the 2020–2121 flood response in China, including command and decision-making, information dissemination, resource mobilization, consultation, emergency rescue, forecasting and warning, and post-disaster resettlement (Jing et al., 2020; Hangzhouwang, 2020), there are limited studies on their functionalities designed for disaster response. Therefore, this study aims to evaluate the functions and usefulness of mobile apps that were available and used during the 2020–2021 Chinese flood seasons, and to explore their potential to promote coproduction, which is defined as "collaborative approaches where citizens or service users engage in partnerships with service professionals in the design and delivery of public service" (OECD, 2011, p. 27).

This study utilizes the Use & Gratifications theory as a conceptual framework and methodology, while the emergency management literature provides theoretical support. The Use & Gratifications theory has been widely used in media

DOI: 10.4324/9781003328896-7

research and emphasizes the distinction between gratifications sought (GS) and gratifications obtained (GO). In this study, the theory provides a basis for evaluating whether the functionality of mobile apps meets the needs of users in disaster situations. As one of the few studies analyzing mobile apps in China from an emergency management perspective, the present research contributes to the study of information and communication technologies (ICT), emergency management, and coproduction in disaster response.

Mobile Apps in Disaster Response

The emergency management literature highlights two mutually reinforcing dimensions of how mobile apps contribute to the disaster response process. First, users expect mobile apps to assist them in performing fundamental emergency management functions, such as providing essential aid or performing critical tasks during an emergency. Second, users expect mobile apps to assist them in acting in specific phases of the emergency management time cycle. These dimensions intersect frequently, but they examine different aspects of the functionality of mobile apps. For instance, an app could assist users in communication for disaster recovery. Communication demonstrates the fundamental functionality of the app, while disaster recovery indicates the specific phase in which it operates.

The 5 C's: Fundamental Functions of Emergency Management in Disaster Response

The fundamental functions of emergency management in disaster response are often summarized using the mnemonic of a few initial Cs. The C3 model, which was originally developed by the military system, has been adopted by the field of crisis and disaster management (Berg, 1990; Williams, 1983). Along with command and control, coordination and communication are also commonly mentioned as key functions (Smith, 2020).

The disastrous response to Hurricane Katrina in 2005 prompted an update of these C models. As a representative, Comfort (2007) advocated for a reframing of emergency management's fundamental functions. She contended that emergency management has traditionally served three purposes: communication, coordination, and control. While the three Cs worked as planned during the government's response to Hurricane Katrina, the emergency management system was unable to conduct contingent and innovative responses to the dynamic circumstances. Comfort indicated the issue as the failure to spur cognition, the awareness of the threat's severity and likely consequences, or the "common operating picture." Cognition has thus been introduced as a core component of the Cs as a result of broad discussion and acceptance of Comfort's argument. To recapitulate, the discipline of emergency management places a premium on five Cs: cognition, communication, coordination (cooperation or collaboration in some contexts), command, and control. These functional concepts can be regarded as what users expect mobile apps to accomplish for emergency management.

- Cognition in emergency management is "a process of continuing inquiry, building on prior knowledge of the region at risk and integrating incoming information on changing conditions and system performance into a current assessment of the vulnerability of the community" (Comfort, 2007, pp. 193–194).
- Communication is a fundamental component of collective action since it entails the transmission of messages between participants and organizations (Kapucu, 2006a, 2006b). Communication also refers to the process of establishing common meanings amongst individuals, organizations, and groups involved in emergencies (Comfort, 2007).
- Coordination is defined as parties cooperating voluntarily to align their actions in order to accomplish a common goal. The majority of studies use the term coordination to describe how organizations collaborate in the aftermath of disasters (Ammann, 2008; Drabek, 1985; Keast & Mandell, 2011; May, 1985; Morris et al., 2007; Tierney, 1985; Oh, 2012).
- Command and control refer to the ability to arrange resources and keep actions focused on the accomplishment of common objectives and to ensure the continuity of operations (Comfort, 2007; United States Joint Chiefs of Staff, 1994). Specifically, command is concerned with the internal direction of an agency's personnel and resources, acting vertically inside the organization. Control places a premium on the overall direction of emergency response actions, acting horizontally across agencies (Berg, 1990).

The Four Phases of Emergency Management

The "four phases" cycle adds another dimension to understanding emergency management. It describes disasters as recurring events with four phases: mitigation, preparedness, response, and recovery. All communities are in at least one phase of emergency management at any time (International Association of Emergency Managers [IAEM], 2008; National Research Council, 1991).

- Mitigation entails actions taken to prevent or reduce the cause, impact, and consequences of disasters.
- Preparedness includes planning, training, and educational activities for events that cannot be mitigated.
- The response phase begins in the immediate aftermath of a disaster, including implementing disaster response plans, conducting search and rescue missions, and taking actions to protect lives and property.
- During the recovery period, restoration efforts occur concurrently with regular operations and activities (National Research Council, 1991).

The four phases circle represents the repetitive stages of the disaster emergency management procedure in its entirety (see Figure 5.1). It assists in identifying for which phase the mobile apps provided support and therefore reflects their capabilities throughout the process.

Figure 5.1 The Four Phases of the Emergency Management Cycle.

Mobile Apps' Functionalities for Disaster Response

The development of mobile apps has revolutionized the way smartphone users cope with disaster situations. Modern mobile apps have become the primary mode of communication, surpassing traditional media such as television or radio (Lindsay, 2011; Wahyu et al., 2012). These apps are capable of providing assistance during emergencies such as road accidents, natural disasters, health crises, and personal safety threats (Anikeeva et al., 2015; Repanovici & Nedelcu, 2021; Sutton et al., 2008; White et al., 2009).

One critical advantage of mobile apps is their ability to facilitate real-time interactions on a large scale, even while on the move. Traditional mass media approaches like radio news rely on a one-to-many communication strategy (Andersen, 2013), which is insufficient for reaching the widest possible audience during disasters (Ghersetti & Odén, 2014). Instead, communication is now shifting to the crowd during disasters (Auferbauer et al., 2015), which means that the public is no longer a passive recipient of information. The crowd can now organize itself, communicate as a network, and provide ongoing assistance to one another during disaster events (Palen et al., 2007). Mobile apps, including mobile social media apps that enable many-to-many mobile interactions, have demonstrated their capabilities in this regard.

Mobile applications enable many-to-many connections, creating opportunities for the coproduction of emergency services. Coproduction refers to collaborative approaches where citizens or service users partner with professionals to design and deliver public services (OECD, 2011, p. 27). It involves planning, designing, delivering, and evaluating public services by drawing on direct input from citizens, service users, and civil society organizations (OECD, 2011, p.32). Elinor Ostrom, who coined the term, described it as a process through which inputs from individuals who are not in the same organization are transformed into goods and services (Ostrom, 1996, p.1073). Coproduction implies that citizens are not passive clients of other sectors but can play an active role in producing public goods and services that are relevant to them (Ostrom, 1996).

Mobile apps have been discovered to empower citizens to report incidents, crowdsource information, and actively participate in rescue operations. Some apps have the ability to connect people and sensors, and share data with the rest of the community (Paletti, 2016).

Building on the literature on emergency management and mobile apps discussed earlier, this study aims to investigate the functionalities and potential for coproduction in China's current flood response. The study will address three research questions:

1. What functions do mobile apps serve in disaster response during the 2020–2021 flood seasons in China?
2. Do the functionalities of the mobile apps align with the key notions of disaster response?
3. Can mobile apps facilitate coproduction of disaster response services?

Uses and Gratifications Framework

This study utilizes the Uses and Gratifications (U&G) Theory as the conceptual framework. U&G theory distinguishes between gratifications sought (GS) and gratifications obtained (GO), which refer to "what users were looking for" and "what they actually received—whether or not it was a fulfilling experience" (Palmgreen et al., 1980). Gratification sought (GS) is defined as the expected benefits from using media, while gratification obtained (GO) is defined as the actual outcomes that users receive (Van de Wiele & Tong, 2014; LaRose & Eastin, 2004).

In this study, GS is interpreted as the desired functionality that helps users carry out the essential functions of emergency management and act in the various phases of the emergency management cycle, as discussed in the literature review. The author examines relevant mobile apps and assesses whether their functionality (which represents the achievable GO) satisfies GS (as depicted in Figure 5.2).

Methodology

This study adopts a field study approach to collect and analyze data. The main objectives are to investigate mobile apps that have functions related to flood emergency management during the 2020–2021 flood seasons. Since some apps may only be available in certain areas, the author selected Zhejiang province to scan mobile apps that serve sub-national levels. Zhejiang province was chosen

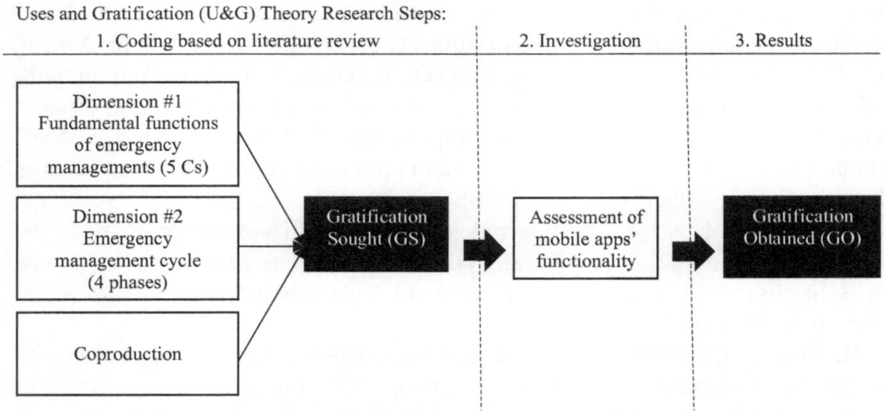

Figure 5.2 A Conceptual Model Based on the Uses and Gratifications Theory.

for three reasons: (1) it is a leader in public affairs mobile apps and mobile Internet development in China (Gu & Lu, 2018; Zheng & Wang, 2019), making it a good reflection of the usefulness of mobile apps in flood emergency management; (2) the private economy and civil society in Zhejiang are relatively well-developed (Yu & Guo, 2012); (3) Zhejiang is a flood-prone province located in the Yangtse River basin and southeast coast, where flood emergency management is a critical public issue.

Steps of Searching and Filtering

Step 1: To find out which mobile apps were used for emergency management purposes, the authors searched for mass news media coverage from June 1, 2020 to September 30, 2021, using the keyword combination "flood"(洪水、洪灾、水灾、防汛、洪涝) + "mobile apps" (移动应用、移动互联网、智能、手机、app) on both Google and Baidu search engines. The search yielded 55 unduplicated news articles mentioning 17 mobile apps with related features.

Step 2: The author searched for the same combination of keywords in the two most popular app platforms, Apple's App Store and Google's Android Market, as well as qimai.cn, a Chinese professional mobile data research platform. The results obtained were then filtered to match the conditions (with the whole country or Zhejiang Province as the service area). The author discovered a total of nine mobile apps in this stage, then deleted those that were duplicated in the previous step and had not been updated since 2020, leaving another five to add to the final list.

Following these two procedures, 22 mobile apps were discovered. Fourteen of them served users nationwide, three served Zhejiang province, and five served Zhejiang province's regional and local levels.

Table 5.1 Functional Purposes of Mobile Apps

Purpose	*Description*
Crowdsourcing	To organize and collect disaster-related data from the crowd
Collaborating platform	To serve as a platform for collaboration during disasters
Alerting and information	To disseminate authorized information before and during disasters
Collating	To gather, filter and analyze data to build situation awareness
Notifying	For users to notify others during disasters
Commanding	Mobilization of personnel and communication of instructions for action

Source: Developed from Tan et al. (2017).

Steps of Analysis

The primary objective of this study is if the functionality provided by the apps satisfies the needs raised from the core notions of disaster response. Due to the lack and fragmentation of data, the actual performance of these apps in reality will only be used as supporting evidence in this study. The analysis will be conducted through the following steps.

Step 1: Functionality examination. The 22 apps were downloaded and examined for their emergency-management-related functions and features. Five properties are examined: types of functions, functional purpose (see Table 5.1), primary functions, features for flood response, and range of service.

Step 2: Check whether the functionality of mobile apps reflects the notions of fundamental functions (five Cs) and the four phases of the disaster response. In other words, the author checks these apps designed to support the function of cognition, communication, coordination, command and control, and the phases of mitigation, preparedness, response, and recovery.

Step 3: Four aspects—assumed user, managed by which sector, developer, and connected sectors—of the apps' functionality were examined to assess their potential in coproduction. Then the author examines the content of news coverage on the related apps to discover additional evidence of how the apps could facilitate coproduction.

Findings

Functionality

Mobile apps used for flood emergency management in China have been discovered to create a highly interconnected, multi-layered, and multi-functional information infrastructure. These apps offer a broad range of robust functions that support flood response. The achievable goals based on their functionality can largely meet users' requirements for emergency management. These mobile apps make extensive use of the latest ICT capabilities, incorporating several innovations. Tables 5.2, 5.3, and 5.4 classify these mobile apps into three categories and provide details on their functionality.

Table 5.2 contains the first group of mobile social media apps that enable people to interact, create, share, and exchange information and ideas in virtual communities and networks. The first group of social media apps mainly facilitates the dissemination and exchange of information about disasters and help-seeking. These apps have special information channels for flood emergencies, which take advantage of their large user base, allowing timely and rich disaster information to circulate quickly among the public.

Table 5.3 contains the second group of tool apps that serve specific purposes such as maps, news, government affairs, and payments. The second group of tool apps leverages their expertise in features tailored for disasters to help users take proactive actions and cope with floods. The interactive nature of the mobile Internet allows these apps to crowdsource professional disaster data during a flood. Map apps collect and publish real-time information on disaster victims, relief resources, and road conditions. Traditional one-to-many news media also experiment with interactive features to play their role as information hubs. For instance, during the 2021 flood season, the CCTV News app opened an emergency mutual aid platform to collect and aggregate help-seeking information and forward it to emergency services. The Paper app also added a feature to gather and publish search, rescue, and help-seeking information.

Table 5.4 contains the third group of apps designed specifically for emergency management. These apps provide various functions and services for flood response, which are detailed in the tables. The third group of built-for-EM apps are professional, full-time emergency management apps that provide robust mobile technical support to professionals and authorities. These apps connect actors across government departments and gather a large amount of disaster information.

These findings can be summarized into three points.

First, the specialized functions and interactivity of these apps allow them to serve and connect various actors during disasters. In flood season, mobile apps including social networks (WeChat), micro-blogging social media (Weibo), online collaboration platforms (Tencent Docs, Shimo, DingTalk), mapping software (Amap, Baidu Map), and government software (Zheliban) can reach and connect a large number of people, information, and public services.

Second, these mobile apps were developed to support multiple levels of actors and organizations. The 14 apps available nationwide are all general-purpose apps and provide a variety of services to a large number of people. They are designed to incorporate flood response measures into their established function design in the event of a disaster. On the other hand, the majority of provincial and sub-provincial apps (7 out of 8) are specifically built for emergency management purposes, aggregating relevant information within their respective jurisdictions and serving as a tool for local responders to contact and act.

From an administrative perspective, these apps cover a wide range of users from nationwide to local jurisdictions. From an operational perspective, they provide

Table 5.2 The Functionality of Mobile Apps (Group 1. Social Media)

App Name	*Type of Purpose*	*Functional Purpose*	*Primary Functions*	*Features for Flood Response*	*Range of Service*
WeChat (微信)	General-purpose	Notifying, Collating, Alerting and information	A multi-media communication mobile app supports single and multiple participation.	Throughout the flood, WeChat was widely and spontaneously utilized to share disaster information among contacts, organize group communication, disseminate news, and perform other emergency management activities.	Nationwide
Weibo (微博)	General-purpose	Notifying, Collating, Alerting and information	Microblogging social media similar to Twitter.	As a large self-media square, Weibo is used to quickly disseminate disaster information during floods.	Nationwide
DingTalk (钉钉防汛)	General-purpose with custom version	Collaborating platform, Commanding	DingTalk is an enterprise communication and collaboration platform.	Through its governmental, Built-for-EM version, members of higher government and lower related departments can communicate without obstacles, and disaster response information can be exchanged among users.	Nationwide
Tik Tok (抖音)	General-purpose	Notifying, Collating, Alerting and information	A short-video social media app.	Users can visually and quickly deliver emergency events and safety situations via short videos of disasters to a large number of strangers.	Nationwide
Kwai (快手)	General-purpose	Notifying, Collating, Alerting and information	Similar to Tik Tok.	Similar to Tik Tok.	Nationwide
Zhihu (知乎)	General-purpose	Alerting and information, Collating	A question-and-answer community similar to Quora.	Volunteer professionals can produce and spread disaster self-help and other types of useful knowledge.	Nationwide

Source: Derived from field study data collected by the author.

Table 5.3 The Functionality of Mobile Apps (Group 2. Tool Apps)

App Name	*Type of Purpose*	*Functional Purpose*	*Main Functions*	*Features for Flood Management*	*Range of Service*
Tencent Docs (腾讯文档)	General-purpose	Collaborating platform, Crowdsourcing	An online document that can be collaborated by multiple users in real time.	Tencent docs were used by volunteers, organizations, and victims to spontaneously exchange information and share self-help guidelines during the floods.	Nationwide
Shimo (石墨文档)	General-purpose	Collaborating platform, Crowdsourcing	Similar to Tencent Docs.	Similar to Tencent Docs.	Nationwide
Alipay (支付宝)	General-purpose	Alerting and information, Crowdsourcing	Alipay is a third-party payment platform and digital life tool.	Alipay helps reduce the need for going out during disasters through online payment services (such as utilities, fees, and fine payment). In Zhejiang, it also serves as a platform for the "Zhejiang Safety Code."	Nationwide
AMap (高德地图)	General-purpose	Alerting and information, Crowdsourcing, Collating	AMap is a free online navigation app.	People can efficiently and accurately navigate through the map to find the nearest disaster shelter and resettlement place, to receive waterlogged spots, and to report help-seeking information.	Nationwide
Baidu Map (百度地图)	General-purpose	Alerting and information, Crowdsourcing, Collating	Similar to AMap.	Similar to AMap.	Nationwide
The Paper (澎湃新闻)	General-purpose	Crowdsourcing, Alerting and information, Collating	A free comprehensive online news website.	Gather and publish search and rescue and help-seeking information after the disaster.	Nationwide

CCTV News (央视新闻)	General-purpose	Crowdsourcing, Alerting and information, Collating	The official client of China Central Radio and Television News.	During the 2021 flood season, an emergency mutual aid platform was opened to collect and aggregate help-seeking information and forward it to emergency services.	Nationwide
Zheliban (浙里办)	General-purpose	Alerting and information, Crowdsourcing	The mobile client of the integrated platform of Zhejiang province government service network.	It centralizes a huge number of government services including emergency, transportation, public security, aviation, and police departments, reducing the need for residents to leave their homes. A flood and typhoon management feature was launched in 2021.	Provincial

Source: Derived from field study data collected by the author.

Table 5.4 The Functionality of Mobile Apps (Group 3. Built-for-Emergency Management Apps)

App Name	*Type of Purpose*	*Functional Purpose*	*Main Functions*	*Features for Flood Management*	*Range of Service*
Zhejiang Safety Code (浙江安全码)	Built-for-EM	Crowdsourcing	A tool built in Alipay and Zheliban to achieve precisely intelligent control and management.	It can report the situation of civilians in a disaster in real-time, greatly reducing the bureaucratic procedures of personnel transfer and rescue, thus achieving precise control of personnel transfer.	Provincial
Flood Control Management (防汛管理)	Built-for-EM	Alerting and information, Collaborating platform, Commanding	Administrative Client of Zhejiang Provincial Headquarters for Typhoon, Flood, and Drought Relief.	The app integrates the weather information and other flood control function modules to achieve a comprehensive presentation of flood control information for decision-making and operation.	Provincial
Ningbo Flood Control (宁波防汛)	Built-for-EM	Alerting and information, Collaborating platform, Commanding	Administrative app of flood control-related departments in the Ningbo area.	This app integrates flood control, water resources, and other functions to provide convenient information and liaison services for relevant personnel of flood control systems.	Regional
Wenzhou Water Conservancy and Flood Control (温州水利防汛)	Built-for-EM	Alerting and information, Collaborating platform, Commanding	Administrative app of flood control-related departments in the Wenzhou area.	This app covers five functional modules: early warning and forecast, information notification, real-time monitoring, basic information, and short-term forecast in the Wenzhou area.	Regional
Jinhua Flood Control Brain (金华防汛大脑)	Built-for-EM	Alerting and information, Collaborating platform, Commanding	Administrative app of flood control-related departments in the Jinhua area.	This app is a comprehensive information platform for Jinhua city-wide real-time disaster information and flood emergency management.	Regional

Tiantai Flood Control PDA (天台县防汛PDA)	Built-for-EM	Alerting and information, Collaborating platform, Commanding	Administrative app of flood control-related departments in the Tiantai area.	The app displays the typhoon path, satellite cloud picture, weather radar picture, and other information in the area on the GIS map. The system assists users in making quick and correct decisions for flood control commands.	Local
Lanxi Mobile Flood Control (兰溪移动防汛)	Built-for-EM	Alerting and information, Collaborating platform, Commanding	Administrative app of flood control-related departments in the Lanxi area.	It provides accurate weather forecasts and real-time warning information to help the Lanxi area protect against floods, typhoons, and other natural disasters in time to avoid unnecessary losses.	Local

Source: Derived from field study data collected by the author.

channels for everyone, from victims to rescuers and administrators, as well as the general public, to become involved in disaster response. Together, these apps offer a powerful and comprehensive set of features that enable a society-wide approach to flood response.

Reflection of Core Notions of Disaster Response by Mobile Apps

By examining whether the functionality of the 22 apps supports the essential functions (the five Cs) and phases of emergency management, this study assesses their potential usefulness in flood response. Table 5.5 demonstrates whether these apps' functionality is designed to support the five Cs of emergency management.

It has been discovered that almost all apps serve the function of promoting cognition. Some of these apps, such as social media and news media, along with the government app Zheliban, provide users with disaster news and warnings. Other apps, such as Dingtalk, Zhejiang Safety Code, and those designed specifically for emergency management, offer crucial organizational and personnel information for effective crisis response. In addition, all emergency management apps provide users with valuable professional information about weather, resources, and more, which can aid in collecting and updating information during a flood response. This greatly enhances users' awareness of the entire disaster situation.

Most apps are designed to facilitate communication in their own unique ways. WeChat, for instance, can be used by any individual user to communicate within their social networks, regardless of the circumstances. Weibo, on the other hand, serves as a platform for the transmission of public information. Online documents are primarily used to communicate between affected people and volunteers. Certain apps also provide channels for help-seeking information and enabling communication between the platform operators and those affected by the disaster. Government and emergency management apps are designed to facilitate communication within responsive organizations.

Furthermore, almost all apps have been discovered to serve the function of facilitating some form of coordination. For instance, there are numerous spontaneous groups on WeChat that form for emergency response purposes. Online documents, government apps, and built-for-EM apps are all platforms that facilitate cooperation and coordination. Apps that have opened channels for help-seeking information have also gained the cooperation of the public in crowdsourcing real-time and valid information, which helps achieve shared goals of protecting lives and property.

The most interesting finding of the study relates to command and control. It was discovered that only government/built-for-EM and related apps have specific functions designed to contribute to command and control during disaster response. This is not solely due to the fact that the responsibility of command is typically assumed by the government. What is noteworthy is that general-purpose apps do not always have a consistent focus on flood response. They often stop maintaining or shut down temporary disaster response functions once the initial crisis has abated. For instance, some nationwide apps opened crowdsourcing features to

Table 5.5 Mobile Apps' Functionality That Supports the Fundamental Functions (Five Cs) of Emergency Management

App Name /Concept	*Cognition*	*Communication*	*Coordination*	*Command &Control*
WeChat (微信)	●	●	●	×
Tencent Docs (腾讯文档)	●	●	●	×
Shimo (石墨文档)	●	●	●	×
Weibo (微博)	●	●	●	×
Baidu (百度)	●	●	●	×
Alipay (支付宝)	●	●	●	×
DingTalk (钉钉防汛)	●	●	●	●
Amap (高德地图)	●	●	●	×
Baidu Map (百度地图)	●	●	●	×
Tik Tok (抖音)	●	●	●	×
Kwai (快手)	●	●	●	×
Zhihu (知乎)	●	●	●	×
The Paper (澎湃新闻)	●	●	●	×
CCTV News (央视新闻客户端)	●	●	●	×
Zheliban (浙里办)	●	●	●	●
Zhejiang Safety Code (浙江安全码)	×	×	×	●
Flood Control Management (防汛管理)	●	●	●	●
Ningbo Flood Control (宁波防汛)	●	●	●	●
Wenzhou Water Conservancy and Flood Control (温州水利防汛)	●	●	●	●
Jinhua Flood Control Brain (金华防汛大脑)	●	●	●	●
Tiantai Flood Control PDA (天台县防汛PDA)	●	●	●	●
Lanxi Mobile Flood Control (兰溪移动防汛)	●	●	●	●

Source: Derived from author's analysis of field study data.

collect and disseminate information about help-seeking and search-and-rescue during the Zhengzhou floods in July. However, based on the author's personal experience with one such app, these temporary features were not always available, and they became unavailable during the Shanxi floods in October 2021. As a result, the flood in Shanxi went unnoticed in the media (VOA, 2021).

Table 5.6 reveals the extent to which apps' functionalities are designed to support the "four phases" of disaster response. The most focused phase by these

Table 5.6 Mobile Apps' Functionality That Supports the Four Phases of Emergency Management

App Name/Phase	*Mitigation*	*Preparedness*	*Response*	*Recovery*
WeChat (微信)	×	•	•	•
Tencent Docs (腾讯文档)	×	×	•	•
Shimo (石墨文档)	×	×	•	•
Weibo (微博)	×	•	•	•
Baidu (百度)	×	•	•	•
Alipay (支付宝)	×	×	×	×
Zhejiang Safety Code (浙江安全码)	×	×	•	×
DingTalk (钉钉防汛)	•	•	•	•
Amap (高德地图)	×	•	•	×
Baidu Map (百度地图)	×	•	•	×
Tik Tok (抖音)	×	×	•	•
Kwai (快手)	×	×	•	•
Zhihu (知乎)	•	•	×	×
The Paper (澎湃新闻)	×	•	•	×
CCTV News (央视新闻客户端)	×	•	•	×
Zheliban (浙里办)	•	•	•	•
Flood Control Management (防汛管理)	•	•	•	•
Ningbo Flood Control (宁波防汛)	•	•	•	•
Wenzhou Water Conservancy and Flood Control (温州水利防汛)	•	•	•	•
Jinhua Flood Control Brain (金华防汛大脑)	•	•	•	•
Tiantai Flood Control PDA (天台县防汛PDA)	•	•	•	•
Lanxi Mobile Flood Control (兰溪移动防汛)	•	•	•	•

Source: Derived from author's analysis of field study data.

apps is "Response," with 20 out of 22 apps providing support in this area. The "Recovery" phase received relatively less support, with only 15 out of 22 apps providing functionality for this phase. Some general-purpose apps open temporary specialized functions when flooding peaks, but do not maintain them until the post-disaster reconstruction and recovery stage. The "Preparedness" phase is supported primarily through the dissemination of warnings and educational messages before a flood is imminent, with 14 out of 22 apps providing support in this area. The phase that has received the least attention is "Mitigation," with only 9 out of 22 apps providing support in this area. Aside from government/built-for-EM apps, only Zhihu has contributed to this phase. As an online question-and-answer community, Zhihu serves as a knowledge repository, making mitigation knowledge provided by volunteer professionals and experts continuously available.

Coproduction in Emergency Management

To reflect on the concept of coproduction based on the functionality of the 22 apps, it is evident that none of the built-for-EM apps engage in cross-sector coproduction. These apps are generally designed as closed systems with restricted access, limited only to necessary public sector personnel, and are owned and managed solely by the public sector. On the other hand, all general-purpose apps are designed to facilitate some form of cross-sector coproduction.

On Tencent Docs, one of the best practice examples occurred (Tencent, 2021). At 20:57 on July 20, the night of day one of the deadly Zhengzhou flood, an ordinary user named "manto" uploaded an open-access document on Tencent Docs titled "information on people who need to be rescued." Following that, many victims and volunteers came in to report, collect, verify, and update information on the real-time situation and to seek support and assistance. The information contained in this document has rapidly developed and adapted to the disaster's changing state. Later, professional nonprofits joined in to organize the information processing. The document has also been updated to include information about government and nonprofit rescue teams. On the document, doctors created online consultation groups and provided medical advice. The technical team at Tencent Docs was on call 24 hours a day to ensure the system's stability. Within 48 hours of its creation, the paper had been viewed by 6.5 million people, edited by over 300,000 people, and thousands of victims had received assistance through it. Moreover, a great deal of information about help-seeking, disaster relief resources, and self-help guidance comes from other social media platforms like WeChat, Weibo, Zhihu, and Tik Tok. Tencent Docs acts as an organizer and filter of information from other sources.

Discussions

In this study, the Use & Gratifications theory provides a conceptual framework for evaluating mobile apps. U&G theory assumes that the audience is active. Blumler (1979) offers several concepts as to the kinds of audience activity in which media consumers could engage. One of the suggested activities is *utility*, which means "using the media to accomplish specific tasks" (Blumber, 1979, p.13). Sundar and Limperos (2013) write that what had been called the "audience" is now referred to as "users," and "usage implies volitional action, not simply passive reception." The findings specified the functionalities of mobile apps in emergency events and thus furthered the U&G theory application in the frontier ICT area and in the field of emergency management.

Answering the Research Questions

The study's findings offer insightful responses to the research questions. Regarding RQ1, the 22 apps analyzed in this study demonstrate that mobile apps can: (1) facilitate communication among multiple users across sectors, including disaster victims and government entities; (2) provide access to information, multimedia communication tools, help-seeking resources, knowledge tips, rescue contacts, and more; (3) enable cross-sector cooperation and coordination, and (4) establish connections between users and service providers during disaster situations through one-to-one, one-to-many, and many-to-many interactions.

In terms of RQ2, the mobile applications have exhibited the capability to support both fundamental functions and all phases of emergency management in general. The results of RQ3 and the real-world case indicate that China's current flood response applications possess the potential to develop into a comprehensive and adaptable information infrastructure in terms of functionality.

Positive outcomes of mobile apps for emergency management also bring two tensions. First, different user groups may seek different gratifications from these apps. General-purpose apps with a large and diverse user base may not consistently provide emergency management functions that meet the expectations of those directly involved in an emergency. As a result, command and control are not significant overall functionality of the apps among the five Cs, and mitigation is the least supported of the four phases. Built-for-EM apps, which are typically professional applications used by the government, can support all five Cs and all four phases. However, these apps are the most closed and isolated and operate within a top-down, bureaucratic model, while general-purpose apps operate within a bottom-up, mass citizen participation model. Collaboration between these two models can help address the weaknesses of each, but this remains a challenge.

The way humans communicate has undergone significant changes over time. Andersen (2013) identified two significant shifts in the media landscape: (1) a transition from top-down to bottom-up interaction, and (2) the emergence of socio-mobile capabilities (Lopatovska & Smiley, 2014). The rise of social media has created a more complex communications environment that is not linear in nature (Andersen, 2013, p. 128). Jenkins' (2006) concept of participatory culture suggests that media technology has the potential to democratize the process of consuming, creating, and distributing content, giving ordinary people greater opportunities to engage in media production. As governments enter the media and communications space through the use of mobile apps, they must consider the unique characteristics of these media and communication platforms. Mobile apps have empowered citizens in various domains, and the public sector should acknowledge and facilitate this reality. Without appropriate management, the information infrastructures of mobile apps will not function effectively. Therefore, it is crucial to establish a philosophy and supporting institutions that promote citizen participation and self-organized emergency management through mobile apps.

The Risks and Limitations of Mobile Apps in Disaster Response

While mobile apps have shown great potential in enhancing emergency management, they also come with certain limitations. One major weakness is their reliance on a robust power and telecommunication infrastructure. In the event of a disaster such as a flood, it is not uncommon for electricity and Internet access to be disrupted, which severely limits the functionality of mobile apps. This was evident during the 2021 flood season in Zhengzhou and neighboring areas.

Furthermore, mobile apps have the potential to widen the digital divide, based on factors such as age, class, language proficiency, and ethnic origin. This means that certain groups may be unable to access or use the app effectively, which could result in unequal distribution of resources and services during an emergency. These limitations must be acknowledged and addressed to ensure that mobile apps are used effectively in emergency management.

Mobile apps pose privacy and information misuse risks despite their rich functionality. Ambiguous privacy policies and non-transparent information management hinder co-production between public and private entities during disaster response efforts. Collaboration and transparent privacy policies are necessary for effective use of private sector-developed apps in disaster response. Developers should prioritize transparency and privacy in their design process, including clear information about data collection and implementation of privacy protection measures. Robust policies are needed to promote reliability, transparency, and citizen empowerment and control over their data.

Conclusion

This study finds that mobile apps have the potential to establish an interconnected, multi-layered, and multi-functional information infrastructure for flood emergency management in China through an analysis of 22 mobile applications that were available or widely used in China during the 2020–2021 flood seasons. These apps provide wide and sophisticated capabilities to aid in the management of flood emergencies.

These apps contribute in their own unique way to emergency management cognition, communication, coordination, and command and control, as well as to the four phases of mitigation, readiness, response, and recovery. Command and control, and mitigation, on the other hand, are rather weak points.

These mobile apps also enable cross-sector collaboration. However, public-sector apps, which have substantial resources and the potential to compensate for the shortcomings of general-purpose apps, appear to be less likely to coproduce. There is a conflict between the present public sector managerial model and the social-empowering nature of mobile apps. This study is limited to its focus on functionality and potential of mobile apps. The performance of apps in the real world is to be further explored by future research.

Note

1 The article is revised based on the article published in *China Media Research* Volume 18, Issue 1, 2022, pp. 38–66. Reproduced with permission from China Media Research.

References

American Red Cross. (2012). *More Americans using mobile apps in emergencies*. Retrieved from: www.prnewswire.com/news-releases/more-americans-using-mobile-apps-in-emergencies-168144726.html

Ammann, W. (2008). Developing a multi-organisational strategy for managing emergencies and disasters. *Journal of Business Continuity & Emergency Planning*, *2*(4), 390–402.

Anderson, A. (2013). *Media, Culture, and the Environment*. Rutgers University Press.

Auferbauer, D., Ganhör, R., & Tellioglu, H. (2015). Moving towards crowd tasking for disaster mitigation. In *The 12th International Conference on Information Systems for Crisis Response & Management (ISCRAM)*, Kristiansand, Norway.

Berg, S. A. (1990). *Introduction to command, control and communications (C3) through comparative case analysis*. Naval Postgraduate School.

Blumler, J. G. (1979). The role of theory in uses and gratifications studies. *Communication Research*, *6*(1), 9–36.

Cheng, E. (2020). *Floods and the coronavirus create more uncertainty for China as food prices climb*. cnbc.com. www.cnbc.com/2020/07/15/floods-coronavirus-create-more-uncertainty-for-china-as-food-prices-climb.html

Cheong, F., & Cheong, C. (2011). Social media data mining: A social network analysis of Tweets during the 2010–2011 Australian floods. *PACIS*, *11*, 46–46.

Comfort, L. K. (2007). Crisis management in hindsight: Cognition, communication, coordination, and control. *Public Administration Review*, *67*, 189–197.

Ding, M. (2022). Mobile apps for flood emergency management in China: Functionality, usefulness, and coproduction. *China Media Research*, *18*(1), 38–66.

Drabek, T. E. (1985). Managing the emergency response. *Public Administration Review*, 45(S1), pp. 85–92.

Ghersetti, M., & Odén, T. A. (2014). Communicating crisis through mass media. In M. Klafft (Ed.), *Current issues in crisis communication and alerting* (pp. 24–41). Fraunhofer Verlag.

Gu, X. & Lu, A. (2018, Jan 22). *浙江基层防汛将实现"痕迹化"管理 [Zhejiang grassroots flood control will achieve "trace" management]*. Xinhua News Network. http://m.xinhuanet.com/2018-01/22/c_1122295799.htm

Hangzhouwang. (2020). *从"数字化防疫"到"数字化防汛" 阿里巴巴助力政府探索数字应急治理 (From "digital epidemic prevention" to "digital flood control" Alibaba helps the government explore digital emergency governance)*. it.hangzhou.com.cn. *Retrieved from:* it.hangzhou.com.cn/jrjd/yjnews/content/2020-07/10/content_7772067.htm

International Association of Emergency Managers. (2008). Principles of emergency management. Retrieved on November 15, 2008 from: www.iaem.com/publications/documents/EMPrinciplesSummary.pdf

Jenkins, H. (2006). *Fans, Bloggers, and Gamers: Exploring participatory culture*. NYU Press.

Jing, L., Yang, Y. H., Zhang, Z. Y., & Jiang, G. (2020) *黑科技 缚洪魔—现代科技提升防汛救灾硬实力 [Technologies to bind the flood devil—modern technology to enhance flood control and disaster relief power]*. xinhuanet.com. www.xinhuanet.com/2020-07/20/c_1126262673.htm

Kapucu, N. (2006a). Interagency communication networks during emergencies: Boundary spanners in multiagency coordination. *The American Review of Public Administration*, *36*(2), 207–225.

Kapucu, N. (2006b). Public-nonprofit partnerships for collective action in dynamic contexts of emergencies. *Public Administration*, *84*(1), 205–220.

Keast, R., & Mandell, M. (2011). The collaborative push: Pushing beyond rhetoric and gaining evidence. In *15th Annual Conference of the International Research Society for Public Management.*

LaRose, R., & Eastin, M. S. (2004). A social cognitive theory of Internet uses and gratifications: Toward a new model of media attendance. *Journal of Broadcasting & Electronic Media*, *48*(3), 358–377.

Latonero, M., & Shklovski, I. (2011). Emergency management, Twitter, and social media evangelism. *International Journal of Information Systems for Crisis Response and Management (IJISCRAM)*, *3*(4), 1–16.

Lindsay, B. R. (2011). Social media and disasters: Current uses, future options, and policy considerations. *Congressional Research Service Report*. Library of Congress. Washington DC.

Lopatovska, I., & Smiley, B. (2013). Proposed model of information behaviour in crisis: The case of Hurricane Sandy. *Information Research: An international electronic journal*, 19(1), n1.

May, P. J. (1985) *Recovering from Catastrophes: Federal disaster relief policy and politics*. Greenwood Press.

Morris, J. C., Morris, E. D., & Jones, D. M. (2007). Reaching for the philosopher's stone: Contingent coordination and the military's response to Hurricane Katrina. *Public Administration Review*, *67*, 94–106.

Ministry of Emergency Management of China. (2021). *应急管理部公布2020年全国十大自然灾害 [The Ministry of Emergency Management announced the top 10 natural disasters in the country in 2020]. China News*. Retrieved from: www.mem.gov.cn/xw/yjglbgzdt/202101/t20210102_376288.shtml

National Research Council (1991). *A Safer Future: Reducing the impacts of natural disasters*. National Academies Press.

Navarro de Corcuera, L., Barbero-Barrera, M. D. M., Campos Hidalgo, A., & Recio Martínez, J. (2021). Assessment of the adequacy of mobile applications for disaster reduction. *Environment, Development and Sustainability*, *24*, 6197–6223.

OECD. (2011). *Together for Better Public Services: Partnering with citizens and civil society*. OECD Publishing.

Oh, N. (2012). Strategic uses of lessons for building collaborative emergency management system: Comparative analysis of hurricane Katrina and hurricane Gustav response systems. *Journal of Homeland Security and Emergency Management*, *9*(1)), 1–28.

Ostrom, E. (1996). Crossing the great divide: coproduction, synergy, and development. *World Development*, *24*(6), 1073–1087.

Palen, L., Hiltz, S. R., & Liu, S. B. (2007). Online forums supporting grassroots participation in emergency preparedness and response. *Communications of the ACM*, *50*(3), 54–58.

Paletti, A. (2016). Co-production through ICT in the Public Sector: when citizens reframe the production of public services. In *Digitally Supported Innovation* (pp. 141–152). Springer.

Palmgreen, P., Wenner, L. A., & Rayburn, J. D. (1980). Relations between gratifications sought and obtained: A study of television news. *Communication Research*, *7*(2), 161–192.

Power, R., Robinson, B., Colton, J., & Cameron, M. (2014). Emergency situation awareness: Twitter case studies. *Information Systems for Crisis Response and Management in Mediterranean Countries,* 218–231. doi:10.1007/978-3-319-11818-5_19

Smith, J. P. (2020, Aug 1). *Fire studies: The four C's of incident mitigation/management* StackPath. www.firehouse.com/leadership/article/21141754/fire-studies-the-four-cs

Sundar, S. S., & Limperos, A. M. (2013). Uses and grats 2.0: New gratifications for new media. *Journal of Broadcasting & Electronic Media*, 57(4), 504–525.

Sutton, J. N., Palen, L., & Shklovski, I. (2008, May). Backchannels on the front lines: Emergency uses of social media in the 2007 Southern California Wildfires. *Proceedings of the 5th International ISCRAM Conference, Washington, DC.*

Tan, M. L., Prasanna, R., Stock, K., Hudson-Doyle, E., Leonard, G., & Johnston, D. (2017). Mobile applications in crisis informatics literature: A systematic review. *International Journal of Disaster Risk Reduction, 24*, 297–311. doi:10.1016/j.ijdrr.2017.06.0

Tencent. (2021, Jul 21). *一个救命文档的24小时 [24 hours of a life-saving document]. Tencent News*. https://xw.qq.com/cmsid/20210721A0F3MP00

United States Joint Chiefs of Staff. (1994). *Department of Defense Dictionary of Military and Associated Terms (Vol. 1)*. Joint Chiefs of Staff.

Van De Wiele, C., & Tong, S. T. (2014, September). Breaking boundaries: The uses & gratifications of Grindr. In *Proceedings of the 2014 ACM International Joint Conference on Pervasive and Ubiquitous Computing* (pp. 619–630).

VOA. (2021, Oct, 10). *山西多日强降雨洪灾严重 信息滞后形成"无人问晋"[Severe flooding caused by several days of heavy rain in Shanxi Province, with a lack of timely information leading to no one caring.].* VOA Chinese. www.voachinese.com/a/china-vast-areas-of-shanxi-flooded-20211010/6265007.html

Wahyu, Y., Shiddiq, M. S. H., & Wahab, M. (2012). Design and realization of an early warning system for natural disaster on digital television in Indonesia. *2012 IEEE International Symposium on Broadband Multimedia Systems and Broadcasting* (BMSB 2012), 1–4.

Williams, P. (1983). Crisis management: The role of command, control and communications. *The RUSI Journal, 128*(4), *33–39.* doi:10.1080/03071848308523493

White, C., Plotnick, L., Kushma, J., Hiltz, S. R., & Turoff, M. (2009). An online social network for emergency management. *International Journal of Emergency Management, 6(3/4), 369.* doi:10.1504/ijem.2009.031572

Yu, J., & Guo, S. (Eds.). (2012). *Civil society and governance in China*. Springer.

Zheng. Y. & Wang, H. (2019). 移动政务的现状，问题及对策 [The current situation of mobile government, problems and countermeasures]. *公共管理与政策评论 [Public Administration and Policy Review*], *8*(2), 73–84.

6 Livestream E-Commerce

The New Social Norm and Its Impact on Chinese Culture

Ping Yang

Introduction

E-commerce (electronic commerce) grows at an exponential pace worldwide and becomes an important trend in practice. With increasing access to mobile communication, more and more people buy their necessary products and services through online shopping (Obrad & Gherhes, 2016). E-commerce is increasingly popular and has a tendency to grow at a fast pace. Pew Research Center (2021) research findings suggest that about 93 percent of American adults use the Internet. About 77 percent of American adults have broadband Internet service at home and 85 percent own a smartphone in 2021. It also notes that roughly 80 percent of Americans are online shoppers, among whom 15 percent of them purchase online on a weekly basis (Smith & Anderson, 2016). With the fast development of e-commerce, studies of its impact on society and culture have become important research issues.

Livestream e-commerce, also called "livestream shopping" or "live commerce," is a trend and digital channel for businesses (Larson, 2021; Benchmark Team, 2022). The US livestreaming market will hit \$25 billion by 2023, according to Coresight Research, a global advisory and research firm (Larson, 2021). Meanwhile, in China, the livestreaming market has grown at breakneck speed, with sales exceeding \$171 billion in 2020 and is predicted to reach \$423 billion in 2022 (Benchmark Team, 2022). By June 2022, the number of Internet users in China reached 1.05 billion, with an Internet penetration rate of 74.4 percent, and users of livestreaming e-commerce have exceeded 716 million in number, accounting for 68.1 percent of the total Internet users, according to the 50th China Statistical Report on Internet Development (CNNIC, 2022). Livestream e-commerce is particularly popular during the COVID-19 pandemic. Even in places where visiting stores is possible, many customers are hesitant to physically go to a store because of health risks. With the rapid growth of livestream e-commerce in China, investigating the new social norm and its impact on Chinese culture has become a contemporary and significant research issue.

The rapid growth of livestream e-commerce in China has ushered in a new research area for scholars in the information and business fields. The role of Chinese culture and the impact of e-commerce on Chinese communication and

DOI: 10.4324/9781003328896-8

society, however, is understudied and lagging behind in this research topic (Fang & Fang, 2022). The current study fills in the void by adding new knowledge of e-commerce by focusing on livestreaming and investigating this new business and social phenomenon from a cultural perspective. This research adds a new lens of study by focusing on the major characteristics of livestream e-commerce that have led to its popularity and success in China, as well as the communication strategies used during the transaction process. An exploration of its impact on Chinese culture in business settings and ethical considerations of business behavior are also significant in contributing to an in-depth understanding of this new social norm. The following section reviews existing literature on electronic commerce and its business models, its development and history in China, the rapid evolution of livestream e-commerce, and traditional Chinese culture and communication, especially in business settings.

Literature Review

E-Commerce and Its Development in China

E-commerce is defined as the buying and selling of goods and services by using computer-mediated electronic networks, such as the Internet (Haag et al., 2005). E-commerce represents the real business model through digitalized applications and computer networks (Guo, 2010). It is associated with the use of online technologies as mechanisms for transferring ownership of the rights to use goods and services (VanHoose, 2011).

Scholars have examined business models during transactions (Aithal, 2016; Laudon & Traver, 2003; Turban et al., 2018). As stated by Aithal (2016), a business model is a set of processes or activities that results in sustainable profit through desired revenue and customer value. Accordingly, an e-business model employs electronic communication technology such as the Internet for exchanging information. The e-business model consists of roles and relationships among customers, firms, allies and suppliers. Laudon and Traver (2003) have listed the business models by type or mode of electronic commerce, as shown in Table 6.1, B2C (business to consumer), B2B (business to business), C2C (consumer to consumer),

Table 6.1 Six Major Business Models in E-Commerce

	Targeted to Consumers	*Targeted to Businesses*	*Targeted to Administration*
Initiated by Business	B2C (business-to-consumer)	B2B (business-to-business)	B2A (business-to-administration)
Initiated by Consumer	C2C (consumer-to-consumer)	C2B (consumer-to-business)	C2A (consumer-to-administration)

Source: Laudon and Traver (2003).

C2B (consumer to business), B2A (business to administration), and C2A (consumer to administration). The above-mentioned six business models are categorized based on whether they are initiated by businesses or consumers, and whether they are targeted to consumers, businesses, or administrations.

E-commerce in China started in 1999 when the website 8848.com was established with a business-to-consumer model (Sun et al., 2019). In the same year, the first consumer-to-consumer business model was adopted by the first Chinese e-commerce company and online trading community EachNet (eachnet.com). EachNet soon became the leading consumer e-commerce platform operator in China because it enabled its users to buy and sell a wide range of items, from clothing and antique calligraphy to computers and real estate, in both auction and fixed-price formats (Sun et al., 2019). As stated by *Global Times*, the e-commerce giant in China, Alibaba (the Alibaba Group), launched Taobao (taobao.com) in 2003, which quickly surpassed EachNet as the market leader in online trading ("E-Commerce Site," 2022). TikTok went live in September 2016 as a music short video creation app by ByteDance (Xu, Yan, & Zhang, 2019). The Chinese version is called Douyin (the Chinese name of TikTok). Douyin started its livestream services in 2017 and immediately cooperated with many businesses to promote sales. Other livestream platforms in China include Kuaishou, Jingdong, Weibo, Pinduoduo, Douyu, and Baidu, while more novel livestream apps are to follow (Sun et al., 2019).

According to China's Ministry of Commerce (2021), national online retail sales in 2020 reached 11.8 trillion Chinese Yuan ($1658 billion), enabling China to become the largest online retail market in the world for eight consecutive years (Department of Electronic Commerce and Informatization, 2021), with the volume of livestream sales exceeding $171 billion in the year 2020 (Larson, 2021). As stated in the livestream e-commerce report by Fang and Yu (2020), the number of livestream e-commerce users in China reached 560 million by March 2020.

Livestream E-commerce

Livestream e-commerce is characterized by its interactivity, synchronicity, and direct communication (Liu, 2021). It differs from the more common formats used for e-commerce (such as pictures and texts) because it incorporates a live video format and is highly interactive (Liu, 2021). Livestream e-commerce has the tendency to go global (Huang et al., 2020). When provided on social media platforms (such as Facebook or Instagram), it is also called social commerce. In addition to social media platforms, livestream e-commerce can be embedded on websites (such as Walmart or Amazon) and apps (such as TikTok/Douyin).

Traditional business transactions in China are conducted with face-to-face interactions between the salesperson and customers on a street market. The nationwide adoption of television in the late 1980s allowed TV shopping as a popular new business platform for consumers. The development of the Internet and computers, and the increasing accessibility to them allowed e-commerce to rapidly grow.

Livestreaming allows influencers to sell products by bolstering credibility and building trust. It is characterized by the fast speed with which the product or service reaches consumers and by the reduction of transaction costs (Obrad & Gherhes, 2016). Its interactivity makes livestreaming an efficient way to learn about products and get questions answered simultaneously (Liu, 2021). The popularity of livestream e-commerce in China attracts consumers of different genders, ages, and educational groups. As previously mentioned, livestream e-commerce in China, between 2017 and 2020, grew at breathtaking speed, with a compound annual growth rate of almost 280 percent (Larson, 2021). This has created new contexts to investigate how technology and communication interact with social and cultural changes. Therefore, the first research question is advanced.

RQ1: What are the characteristics of livestream e-commerce that contribute to its popularity in China?

Chinese Cultural Traditions

To understand the communication behaviors of e-commerce and its impact on Chinese culture, it is necessary to understand the fundamental components of that culture. The contemporary conceptualizations of Chinese culture are deeply rooted in Chinese philosophical traditions of Taoism, Buddhism, and Confucianism (Nakayama & Martin, 1998). Influenced by these cultural forces, Chinese culture has been categorized as high context.

High-Context Culture. Hall's (1966) ground-breaking conceptualization of culture as high-context (HC) versus low-context (LC) has become the fundamental starting point for much research exploring issues of culture and communication (Chuang & Hale, 2002). Chinese culture has been considered near the high-context culture end (Chuang & Hale, 2002). Communication in high-context culture can be described as occurring in degrees of context, with people from Asia, including China, usually classified as a higher context culture as opposed to people from North America, who are classified as a lower context culture (Hofstede, 1980). Brislin (1993) has noted that in a high-context culture, understanding is implicit and comes not so much from the words themselves but from meanings surrounding the situation, environment, and nonverbal behaviors.

Indirect Communication. The Chinese culture and communication are interrelated. According to Gao (1997), the primary functions of communication in Chinese culture are to maintain existing relationships among individuals, to reinforce role and status differences, and to preserve harmony within the group. Based on cross-cultural and intercultural comparisons, research findings have revealed indirect communication patterns in Chinese culture. The indirect communication style is in accordance with the Chinese philosophical principles of limited use of verbal communication, while emphasizing the nonverbal messages and contextual clues in understanding the transmitted information in conversations. According to Gudykunst and Ting-Toomey (2003), in the indirect communication style, the verbal message is often so designed as to camouflage the speaker's true intentions, needs, wants, and desires. People of East Asian cultures generally use an indirect mode of communication, whereas people of North American culture predominantly use a direct mode of communication.

Online communication has provided a unique context to study culture and has changed the cultural landscape of people (Yang, 2023). Scholars have examined how culture presents different characteristics in the cyberspace regarding anonymity, synchronicity, identity, and values (Caparas & Gustilo, 2017; Yang, 2020; Shuter et al., 2018). Yang's research (2012) has focused on cultural differences in online communication and stated that a culture can be both high-context and low-context. Chinese culture and communication on TikTok, for example, can be low-context and direct because of its emphasis on explicit expression of thoughts and ideas (Yang, 2022). Chinese people, especially the younger generations, are more direct in showing their emotions and desires during online interactions. Therefore, it is important to investigate the changing culture because of the growing livestream e-commerce in Chinese society. The following section reviews existing literature on cultural values in business settings.

Cultural Values in Business

Business conduct evolves around the philosophical, societal, economic, religious, and institutional concepts and notions of culture (Crane & Matten, 2004; Scholtens & Dam, 2007). The countries of the world mainly fall under two business cultures: (1) formal versus informal culture and (2) relationship-focused versus deal-focused culture (Gesteland, 1999). Countries with a formal business culture include most countries in Europe, Asia, the Mediterranean region, Latin America, and the Arab world. The informal business culture includes Australia, Denmark, the United States, Canada, Norway, Sweden, and New Zealand (Gestland, 1999). According to Zhu (2016), Chinese business culture is formal and relationship-focused.

Formal Business Culture. The difference between a formal and informal culture is that the formal culture tends to be organized in steep hierarchies that reflect major differences in status and power. According to Zhu (2016), formal business culture is concerned with how serious and formal a behavior of business must be to the owner or employer. In other words, everything concerning business has to be formal: words used, dress, social courtesy, formality, locations of meetings, salutations, respect, and so forth. An informal culture values more egalitarian organizations with smaller differences in status and power. In the informal business culture, hierarchy is absent. Addressing each other is done in a casual way, with few ceremonies, procedures, feasts, or gift exchanges. As Gestland (1999) noted, Chinese people conduct business mainly by following the formal etiquette and rules. Everything is formal or ceremonious.

Relationship-Focused Culture. Countries with a relationship-focused culture are characterized by reluctance to do business with strangers, making initial contact indirectly, and using meetings to build up trust before getting down to business (Gestland, 1999). They tend to emphasize the maintenance of harmony, dislike conflict and confrontation, prefer indirect and high-context communication, and rely on close relationships rather than contracts to resolve disagreements (Gestland, 1999; Zhu, 2016). As noted by Gestland (1999), many Asian countries like China, Japan, Korea, Philippines, Singapore, and India belong to relationship-based cultures. In contrast, deal-focused cultures are characterized by their openness to do business with strangers, willingness to contact potential customers or business partners directly, and getting down to business at meetings after a few minutes of general conversation (Gestland, 1999; Zhu, 2016). Countries such as the United States, France, Britain, Germany, and many northern European and Northern American countries are deal-based cultures because they are more willing to use direct, frank, low-context communication style and rely on written agreements rather than personal relationships (Gestland, 1999).

Guanxi and Rhetorical Strategies. Additionally, in the Chinese community Guanxi is strongly emphasized in doing business. Guanxi can be defined as the special relationships two persons have with each other (Sheh, 2001). It refers to the use of business networks, personal relationships, and connections in doing business (Li, 2008). In Chinese business networking, cordial guanxi is a necessary connection in ensuring such networks operate reliably (Sheh, 2001). To be persuasive, three rhetorical strategies have been identified in communication: ethos, pathos, and logos (Braet, 1992). Ethos is "a Greek word that is closely related to our terms ethical and ethnic" (Campbell & Huxman, 2009, p. 232). It refers to the knowledge of characters. Aristotle discussed logos as the logical means of proving an argument (Braet, 1992). Logos is based on logical reasoning. In addition, the speakers must rely on pathos, the strong role that emotions play in effectively persuading the audience. This research focuses on the unique context of e-commerce in examining the communication behaviors and rhetorical strategies in doing livestream business online.

Livestream e-commerce offers a new way of doing business and creates a new social phenomenon. A plurality of experts believe that the new norm of the world will be more tech-driven and "tele-everything" (Anderson et al., 2021). People's relationship with technology will deepen as they rely more on digital connections for daily commercial transactions, social interactions, and life activities. The communication patterns and cultural values in online e-commerce, especially livestream shopping, provide a unique context to be examined. Although livestream e-commerce is flourishing in China and growing fast in many parts of the world, few research studies have explored its popularity and impact on society and culture. The current research project provides a unique perspective in studying the intersection of technological innovations and new social norms by putting forward two research questions:

RQ2: What communication strategies contribute to the success of livestream e-commerce in China?

RQ3: What is the impact of livestream e-commerce on the changing cultural practices in China?

Research Method

To examine livestream e-commerce in China, Taobao and Douyin have been selected for data collection due to their large number of active users. Taobao is the earliest e-commerce platform established by China's e-commerce titan Alibaba in 2003. It features products and services on both the website www.taobao.com and through the Mobile Taobao App. It is China's largest mobile commerce platform, with 710 million active users according to iiMedia Research (2021). Douyin, also named TikTok for global users, soon after being launched in 2016 by byteDance became one of the most frequently used short video platforms in China (Xu et al., 2019). With 680 million active users and 70.9 percent of the population being current or previous users, TikTok ranked as top among the short video apps in 2020 (iiMedia Research, 2021). A qualitative content analysis of the two livestream shopping giants has enabled the researcher to conduct an in-depth understanding of the most popular business activities in China's livestream e-commerce field.

Qualitative content analysis is a widely used research technique for interpreting meanings from content (Hsieh & Shannon, 2005). It is the best fit for the current research project because of its focus on a systematic analysis of textual, visual, and audio communication behaviors based on qualitative-interpretive steps. The researcher has viewed and analyzed livestream e-commerce activities on the two platforms over a period of three months, from July to September 2022. A total of 56 livestream shopping rooms have been visited by the researcher to collect data, half from Taobao and half from Douyin. These shopping rooms were selected because of their popularity, each exceeding 10,000 viewers per day during the three months, as indicated on the platforms. The researcher used a pseudonym when visiting these sites, participating in conversations and taking notes. Coding was conducted based on notes and observations of three units: characteristics, communication strategies, and cultural impact. No issues with anonymity or confidentiality occurred during the data collection process.

Accurate, authentic, and adequate research data guaranteed the relativity and validity of the qualitative research process. Repeated analysis and readings of the data allowed the researcher to secure the findings of the studied phenomenon. A worthy topic, in-depth understanding, rigorous interpretation, and the ethics of research are all criteria to ensure the quality of the current qualitative research (Croucher & Cronn-Mills, 2018). The following section presents research findings regarding the characteristics of livestream e-commerce, which lead to its popularity and success. Communication strategies allow the livestream hosts to effectively interact and persuade the consumers to buy their products. Also, the research question regarding its impact on culture is investigated through low-context and deal-based values. With more and more people using mobile communication to engage in online shopping activities, the research findings are of great significance in understanding this new norm and its impact on contemporary Chinese society and culture.

Analysis and Research Findings

Characteristics Contributing to the Popularity of Livestream E-commerce

The analysis of the livestream e-commerce on Taobao and Douyin has revealed four major characteristics of livestream e-commerce that contribute to its popularity among Internet users in China. The first features are the product quality and lower price.

Product Quality and Lower Price. One of the defining characteristics of livestream products is the quality of products and their appealing prices. No matter if the sellers use the C2C or B2C model, they reduce the prices and offer discounts to allure potential buyers. The sellers repeatedly mention how they offer a lower price compared to other sellers and how good the quality of their product is. As Shi stated (2020), three significant factors influence the consumer's decision-making process: the hosts, the products, and the platforms. The majority of the sellers try to attract customers and persuade them to make a purchase by emphasizing the superior quality of their merchandise and how, compared to other sellers, their price is more reasonable and worth buying.

Easy-to-Use Technology and Services. Both Taobao and Douyin offer convenient services on their short-video platforms. The audience can easily watch the demonstration and try-on for clothing; they can also view the price and other related products simultaneously. While listening to the hosts, who are displaying the products and explaining, the consumers can also make use of other links and services to further explore the site. For example, at the top of the video there is information on "the number of current viewers" and "follows." At the bottom of the video, there is a dialogue box for customers to ask questions and leave comments. There are also links to "show more products," "interact with the seller," "click likes," and "show hearts," as well as links for "more information on products" and to "add to shopping cart." From the consumers' perspectives, the multiplicity of these links and services makes them easy to use, learn, follow, and pay for.

Synchroneity and Delivery Time. The chat box at the bottom of the screen allows spontaneity and synchroneity when interacting with the seller. The demonstration is a good technique being used to show the products. The sellers continuously try different clothes, cosmetic products, kitchen utensils, electronic devices, jewelry, food, furniture, and other articles for daily use. What is also attractive is that the platforms of Taobao and Douyin offer immediate delivery services. They have promised that most products arrive within the same day or the next day or two to customers within the country.

Feedback and Return Policies. Even if the livestream stops during the night, recorded videos about products are available on the site for customers who are interested in continuing to watch. Additionally, times are posted with regard to when the livestream introduction will start again the next morning. Warm messages such as "welcome to my livestream room" or "hello, new and returning customers" are still played along with the recorded video. In addition to the "likes" and "follows" on the main screen, the links for interaction also guide customers to "copy the link of this product," "contact the seller via 'WeChat' or 'Text,'" as well as "scan the code for this product" are available for consumers to contact the sellers even during their absence. Return policies are clearly stated on the payment page, usually seven days after the purchase date.

Communication Strategies

The research findings have also shown five communication strategies employed during online business transactions, including communication styles and the language used by the hosts in being persuasive and satisfying the needs of customers.

Dialogic and Direct Communication. The livestream hosts interact with the audience all the time by providing detailed information about their products, asking and answering questions, inviting comments, feedback, and sharing their experiences. Dialogic communication is one of the features of online commerce because all of the communication is informal, natural, and casual. No strict business netiquettes are required. Frequent use of humor and teasing have been used, resulting in a good laugh between hosts and consumers. Messages are explicitly delivered and understood by the viewers.

Multiple Channels and Stimulation. Some livestream hosts are dressed in casual clothes, some are in traditional dress, or semi-casual business attire while talking with the customers. The livestream video rooms are all well-designed and prepared, with the products displayed in the center. Some livestream rooms have the brand name of their products or a slogan displayed at the top of the screen. Additional stimuli include background music, pleasant colors, room setups, products stage, and easy-to-detect coupons. Audio and video stimuli are constantly displayed for customers to have a better shopping experience.

Persuasiveness and Language Use. The language used during livestream e-commerce is generally simple, conversational, natural, spontaneous, precise, and inclusive. The sellers constantly use words such as honey (亲), dear (亲们), babe (宝贝), sisters (姐妹们), and brothers (兄弟们) when addressing customers to increase the intimacy of their connections with the viewers. Additionally, stylized language such as repetition, exaggeration, and alliteration are frequently used to persuade customers. Furthermore, the livestream hosts are careful in using only inclusive language and avoiding profane and exclusive language when they talk with potential buyers. They try to sound like good friends who know of a nice product and would recommend it because of these close relationships.

Rhetorical Strategies and Audience-Centered Approach. The three rhetorical strategies emphasized by Aristotle, namely ethos, pathos, and logos (Braet, 1992), have all been identified during transactions. The livestream hosts have done preparation ahead of time, acquiring sufficient knowledge about the products, preparing for potential questions from the viewers, calculating how much consumers might have saved, arguing why this product is a must-buy, and making use of emotions to influence the buyers. Therefore, all three rhetorical strategies of ethos, pathos, and logos have been shown during the process. Guanxi, however, is rarely a factor in doing business in livestream e-commerce. With over 10,000 viewers in the host room to watch and to make conversation, the communication mainly emphasizes the needs and wants of the audience. It is audience-centered, with the goal of completing transactions instead of focusing on personal relationships or business networks.

Satisfying the Needs for Socializing and Interacting. According to Will Schutz's three-dimensional theory of interpersonal behavior (Schutz, 1966), human beings need to communicate with each other to satisfy our interpersonal needs, including (1) the need for control; (2) the need for inclusion; and (3) the need for affection. We have the motivation to maintain some degree of influence during communication, the need to be included in some activities, and the need for others to appreciate us and show their affection to us (Schutz, 1966). Livestream e-commerce fulfills these needs during transactions. Statistics show that the female audience accounts for 68.8 percent of Taobao viewers (Fang & Yu, 2020). They have the need for socializing, interacting, and communicating with others. During livestream e-commerce, people are no longer passive buyers browsing through webpages or items on the shelves – they are real people engaging in talking, listening, questioning, answering, and exhibiting their sense of humor. Purchase is an activity that can be combined with our needs for socializing, relationships, and learning. It meets our needs for interpersonal behaviors so that returning customers come back time and again to make more purchases of the products.

Impact on Culture

To address the third research question on the impact of livestream e-commerce on Chinese culture in business settings, research findings have indicated a low-context and deal-focused culture, in contrast to the traditional high-context and relationship-focused culture. The cultural changes were demonstrated through online communication in livestream host rooms but may as well make a significant impact on business culture offline and contemporary Chinese culture in the long run.

Low-Context Culture and Directness. Livestream e-commerce has revealed the change from the traditional high-context culture to low-context in the online livestream communication context. The hosts are communicating explicitly about their products, sharing information about price changes, special offers, and discounts by using a direct way to communicate with the customers. They often urge the viewers to click the links at the bottom of the page to put the product in their shopping cart, to "like" or "follow" their videos, or simply to click the "buy now" button immediately.

Informal and Deal-Focused Culture. The culture in livestream commerce is informal. Most of the hosts use a conversational style to communicate with the audience. They use intimate words to address the customers. It is a more deal-focused culture in livestream e-commerce. The long-term relationship between businessmen and customers is not a concern on the site. Instead, short-term completions of deals (as many as possible) become the focus during conversations. Market competition and economic variables such as price and quality outperform the Guanxi type of business practices in online e-commerce.

Individualism and Self-Fulfillment. A person's self-concept is influenced by not only themselves, but also by culture, media, and the surrounding environment. It is through one's reflected appraisal of how others think of oneself, that one understands who he/she is (Floyd, 2021). Additionally, social comparison also allows us to observe how we compare with others in order to evaluate our characteristics and form our identities (Floyd, 2021). In livestream e-commerce, the hosts emphasize self-concept and individual accomplishments, as shown by the sales amount by the end of the day. When many hosts become rich overnight, it is definitely a huge impetus to urge people, especially the youth, to dream of becoming a livestream host to successfully sell products and become rich within a short period of time. Livestream e-commerce appears as a convenient way for self-fulfillment and to achieve personal wealth.

Participatory Culture and Self-Expression. Participatory culture is a significant phenomenon in livestream e-commerce. Both hosts and viewers are active participants in business activities, with enthusiasm and excitement. The hosts are eager to promote their sales, while the consumers hope to learn more and pick the best merchandise. It is a new way of participation, a form of democracy, and a method of self-expression. Everyone feels they are on a level field playing the game because they are active participants. Social hierarchies in reality are giving way to the seemingly fair game of online shopping. For people who live in a culture that emphasizes indirectness and social hierarchies, being able to participate and self-express is highly valued and appreciated. This is demonstrated through the rapid language exchanges, direct communication, high enthusiasm, and the urge to voice their opinions and negotiate the prices right at the moment.

Livestream Celebrities and Their Influence. More and more micro-celebrities are emerging from livestream e-commerce. They have gained fame, wealth, and influence. According to the *Harvard Business Review* ("Livestream E-Commerce," 2022), top livestream hosts such as Weiya and Li Jiaqi, respectively, received 3.4 and 3.8 billion Chinese Yuan ($480–530 million) from January to March 2020. Television celebrity Liu Tao gained a GMV (gross merchandise value) of 220 million Chinese Yuan ($31 million) on a single day in June 2022. Similarly, Huang Shengyi and Yang Zi reached a GMV of 128 million Chinese Yuan ($18 million) on one day in June 2022. The lines between self-made stars in livestreaming and celebrities in reality are becoming more and more blurred. They have interchangeably made use of their influence to sell products, accumulate power, and achieve economic success.

Impact on Local and National Economies. Livestream e-commerce has undoubtedly made a tremendous impact on local and national economies. More and more people from all fields of life, including leaders of companies, government officials, and farmers from remote areas have joined livestream e-commerce to become hosts. The BBC news reported that in order to boost the local economy in Hunan Province, a vice mayor in Hunan Province became a micro-celebrity on a livestream while selling local tea from his city (Wang, 2020). In April of the same year, Luo Yonghao, a Chinese entrepreneur and the chief executive officer and founder of the Chinese technology company, Smartisan, became a micro-celebrity by within three hours reaching a GMV of 110 million Chinese Yuan ($15.5 million), with over 4.9 million viewers. This was the top GMV record in Douyin. The rapid increase of livestream e-commerce has made a tremendous impact on the local and national economies and is likely to continue growing in the years to come.

Ethical Considerations and Future Directions

Ethical Considerations. Many people have raised concerns over ethical business behaviors during livestream e-commerce. Problems with trust are constantly pointed out by online customers. Since the COVID-19 pandemic, many hosts have experienced what is called "frequent rollover," which means they have been punished by being unfaithful to customers, engaging in cheating behaviors, or selling products with questionable quality. This problem has caused damage to the reputation of a few livestream stars and celebrities. Some even fall into legal accusations. From the consumers' perspective, meeting customers' expectations about product quality and making delivery time real are key to establishing ethical relationships with customers.

Fierce and Disorganized Competition. Many livestream hosts boast of their merchandise price and claim to offer the lowest price possible. This has led to fierce competition among hosts who sell similar or related products. Some hosts have lost the trust of customers or forced customers to take legal action. Consumers also have begun to challenge the quality of their products, and this has reduced their trust. This makes the livestream market less organized and requires better policies for fair competition and customer protection. It is important to establish an orderly and healthy competitive environment in livestream e-commerce. What is more, the over-reduction of the price will do potential harm to the manufacturers, hindering them from increasing production and improving product quality.

During the Pandemic and In the Future. Livestream e-commerce has grown exponentially during the pandemic and become an important business market in China. COVID-19 has had a profound impact on practically every element of our lives. The pandemic shifted how we connect daily, even with mundane chores like shopping. More and more consumers are focusing on livestream e-commerce even as stores open back up. According to Keane and Yu (2019), the slogan "cybercommunity of shared destiny" was unveiled as the theme of the 2016 World Internet Forum in Wuzhen, China, to indicate "properly managed cyberspace." Whether livestream e-commerce will remain as popular after the pandemic and in which direction it goes remains a significant topic for people to discuss, research, and analyze for now and in the future.

Conclusion

This research has investigated the livestream e-commerce of the two largest Chinese e-commerce giants, Taobao and Douyin. The research findings have shown four main characteristics of this new business model in its growing popularity. It has been attracting millions of viewers and making great success due to its high product quality and lower price, easy-to-use technology and user-friendly services, synchroneity, and fast delivery time, as well as favorable feedback and return policies.

Additionally, five communication strategies have been unraveled from the research findings to present ways that hosts try to persuade the audience and make transactions. Their communication strategies are different from the traditional Chinese communication patterns by being more dialogic and direct, using multiple channels and stimulation, resorting to a variety of language styles and persuasive words, using rhetorical strategies and being audience-centered, as well as focusing on satisfying the viewers' needs for socializing and interacting through their conversations. The impact of livestream e-commerce on Chinese culture in business settings has been revealed through six perspectives, including low-context culture and directness, informal and deal-focused culture, individualism and self-fulfillment, participatory culture and self-expression, livestream celebrities and their power, as well as its influences on local and national economies. What is more, the ethical considerations and future directions of livestream e-commerce in China have been examined and discussed.

With the rapid growth of e-commerce worldwide, livestream e-commerce in China has been increasing at a quick rate, especially during the pandemic. With more and more people using mobile communication to engage in online activities such as livestream shopping, the research findings are of great significance in understanding this new social norm, its characteristics, and its impact on contemporary Chinese cultural and communication practices.

References

Aithal, P. S. (2016). A review on various e-business and m-business models & research opportunities. *International Journal of Management in Education, 6*(1), 275–298.

Anderson, J., Rainie, L., & Vogels, E. A. (2021). *Experts say the 'new normal' in 2025 will be far more tech-driven, presenting more big challenges*. Retrieved from: www.pewresearch.org/internet/2021/02/18/experts-say-the-new-normal-in-2025-will-be-far-more-tech-driven-presenting-more-big-challenges/

Benchmark Team. (2022). *Is livestream shopping the future of e-commerce?* Retrieved from www.benchmarkemail.com/blog/livestream-shopping-ecommerce/

Braet, A. C. (1992). Ethos, pathos and logos in Aristotle's rhetoric: A re-examination. *Argumentation, 6,* 307–320.

Brislin, R. (1993). *Understanding culture's influence on behavior*. Harcourt Brace College.

Campbell, K. K., & Huxman, S. S. (2009). *The rhetorical act: Thinking, speaking, and writing critically*. Wadsworth.

Caparas, P., & Gustilo, L. (2017). Communicative aspects of multilingual code switching in computer-mediated communication. *Indonesian Journal of Applied Linguistics, 7*(2), 349–359.

China's Ministry of Commerce (Nov. 1, 2021). *New chapter of e-commerce development with high quality.* Retrieved from www.gov.cn/zhengce/2021-11/01/content_5648151.htm

Chuang, T., & Hale, C. (2002). A critical examination of the Eurocentric representation of Chinese communication. In W. Jia, X. Lu, & D. Heisey (Eds.), *Chinese communication: Theory and Research: Reflections, new frontiers, and new directions* (pp. 67–84). Ablex Publishing.

CNNIC (China Internet Network Information Center). (2022). *The 50th China statistical report on internet development.* Retrieved from: www.cnnic.cn/n4/2022/0914/c88-10226.html

Crane, A., & Matten, D. (2004). *Business ethics.* Oxford University Press.

Croucher, S. M., & Cronn-Mills, D. (2018). *Understanding communication research methods: A theoretical and practical approach* (2nd ed.). Routledge.

Department of Electronic Commerce and Informization (January 24, 2021). MOFCOM (Ministry of Commerce of People's Republic of China) Department of Electronic Commerce and Informatization on the development of the online retail market in 2020. Retrieved from: http://english.mofcom.gov.cn/article/newsrelease/policyreleasing/202101/20210103034679.shtml

E-commerce site eachnet.com, once owned by eBay, ceases operation due to fierce competition (2022). *Global Times*. Retrieved from www.globaltimes.cn/page/202207/1271244.shtml

Fang, G., & Yu, Q. (2020). *Livestream e-commerce report MCN: Gather new users to boost new economy*. Retrieved from: http://pdf.dfcfw.com/pdf/H3_AP202006181385705779_1.pdf

Fang, Y. & Fang, L. (2022). A review of Chinese e-commerce research 2001-2020. *International Journal of E-Business Research (IJEBR), 18*(1), 1–22.

Floyd, K. (2021). *Interpersonal communication* (4th ed.). McGraw-Hill.

Gao, G. (1997). "Don't take my word for it." – Understanding Chinese speaking practices. *International Journal of Intercultural Relations, 22,* 163–186.

Gestland, R. (1999). *Cross-cultural business behavior: Marketing, negotiating, and managing across cultures*. Copenhagen Business School Press.

Gudykunst, W., & Ting-Toomey, S. (2003). *Communicating with strangers: An approach to intercultural communication* (4th ed.). McGraw-Hill.

Guo, C. (2010). The psychological influences of threat on B2C e-commerce adoption: An extended study and explanatory framework. *Journal of Promotion Management, 16,* 303–315.

Haag, S., Baltzan, P., & Phillips, A. (2005). *Business driven technology*. McGraw-Hill/Irwin.

Hall, E. (1966). *The hidden dimension*. Doubleday.

Hofstede, G. (1980). *Culture's consequences: International differences in work-related values*. Sage.

Hsieh, H., & Shannon, S. E. (2005). Three approaches to qualitative content analysis. *Qualitative Health Research, 15*(9), 1277–1288.

Huang, H., Blommaert, J., & Praet, E. V. (2020). "Oh, my God! Buy it!" A multimodal discourse analysis of the discursive strategies used by Chinese ecommerce live-streamer Austin Li. *HCI International 2020 (International Conference on Human-Computer Interaction) – Late Breaking Papers: Interaction, Knowledge and Social Media,* 305–327.

iiMedia Research (2021). *The goal is 680 million DAUs on Douyin in 2021: Development analysis of Chinese short video platforms.* Retrieved from: www.iimedia.cn/c1020/78071.html

Keane, M., & Yu, H. (2019). A digital empire in the making: China's outbound digital platforms. *International Journal of Communication, 13,* 4624–4641.

Larson, K. (2021). Retailers embrace livestreaming, market expected to reach $11 billion In 2021. Retrieved from www.forbes.com/sites/kristinlarson/2021/03/27/retailers-embrace-livestreaming-market-expected-to-reach-11-billion-in-2021/?sh=2f12409c2fde

Laudon, K. C., & Traver, C. G. (2003). *E-Commerce: Business, technology, society.* Addison- Wesley.

Li, M. (2008). When in China. *Communication World, 25*(6), 34–37.

Liu, F. (2021). *Livestream ecommerce: What we can learn from China.* Retrieved from: www.nngroup.com/articles/livestream-ecommerce-china/

Livestream e-commerce: Short-lived or long-lasting? (2022). *Harvard Business Review.* Retrieved from https://m.hbrchina.org/article/239

Nakayama, T., & Martin, J. (1998). *Worldviews, religion, and intercultural communication.* In J. Martin & T. Nakayama (eds.), *Readings in intercultural communication: Experiences and contexts* (pp. 26–35). McGill-Hill Companies.

Obrad, C., & Gherhes, V. (2016). Attitudes towards online commerce: A case study of student population in Timisoara. *Professional Communication and Translation Studies, 9*, 47–52.

Pew Research Center (June 2021). *Mobile technology and home broadband 2021.* Retrieved from www.pewresearch.org/internet/2021/06/03/mobile-technology-and-home-broadband-2021/

Scholtens, B., & Dam, L. (2007). Cultural values and international differences in business ethics. *Journal of Business Ethics, 75*, 273–284.

Schutz, W. (1966). *The interpersonal underworld.* Science and Behavior Books.

Sheh, S. W. (2001). Chinese cultural values and their implication to Chinese management. *Singapore Management Review, 23*(2), 75–83.

Shi, L. C. (2020). *Three factors for livestream ecommerce: hosts, products, and platforms.* Retrieved from www.woshipm.com/it/3550549.html

Shuter, R., Dutta, U., Cheong, P., Chen, Y., & Shuter, J. (2018). Digital behavior of university students in India and the U.S.: Cultural values and communication technologies in the classroom. *Western Journal of Communication, 82*(2), 160–180.

Smith, A., & Anderson, M. (2016). *Online shopping and e-commerce.* Retrieved from: www.pewresearch.org/internet/2016/12/19/online-shopping-and-e- commerce/

Sun, B., Ouyang, R., Ju, X., He, Y., & Chang, Y. (2019). *The waves are more powerful: Twenty-year development of e-commerce in China.* China Internet Economy Research Institute. Retrieved from: www.d-long.com/eWebEditor/uploadfile/20190627103303159588887.pdf

Turban, E., Outland, J., King, D., Lee, J. K., Liang, T., & Turban, D. C. (2018). *Electronic commerce: A managerial and social networks perspective.* Springer.

VanHoose, D. (2011). *Ecommerce economics*. Routledge.

Wang, Y. (April 2020). Livestream e-commerce: New trend during pandemic. *BBC News.* Retrieved from www.bbc.com/zhongwen/simp/chinese-news-52392688

Xu, L., Yan, X., & Zhang, Z. (2019). Research on the causes of the "TikTok" app becoming popular and the existing problems. *Journal of Advanced Management Science, 7*(2), 59–63.

Yang, P. (2012). Knowing through asynchronous time and space: A phenomenological study of cultural differences in online interaction. In K. St. Amant & S. Kelsey (eds.), *Computer-mediated communication across cultures: International interactions in online environments* (pp. 108–122). IGI Global.

Yang, P. (2020). Reconceptualizing intercultural competence in a networked society. *China Media Research, 16*(2), 54–65.

Yang, P. (2022). TikTok and internet celebrity: An analysis of the impact of short video apps on Chinese culture and society. *China Media Research, 18*(1), 23–37.

Yang, P. (2023). Understanding intercultural conflict in virtual space: Characteristics, approaches, and management styles. In X. Dai & G. M. Chen (eds.), *Conflict Management and Intercultural Communication: The Art of Intercultural Harmony* (2nd ed., pp. 330–342). Routledge.

Zhu, P. (2016). Impact of business cultural values on web homepage design that may hinder international business. *Journal of Technical Writing and Communication, 46*(1), 105–124.

7 How Do Mental Health Non-Governmental Organizations (NGOs) in India Use Instagram?

Roma Subramanian

Introduction

Mental illness is a leading contributor to the disease burden in India (Sagar et al., 2019). The World Health Organization (2020) estimates roughly 7.5 percent of India's 1.3 billion people suffer from a mental illness or disorder. In 2017, the prevalence of mental disorders in India was 1 in 7 (Sagar et al., 2019). Furthermore, India has one of the highest suicide rates in Southeast Asia (Dandona et al., 2018). Experts predict that these numbers will only increase in the upcoming decades as COVID-19 exacerbates the existing mental health crisis (Sharma et al., 2020). The mental health crisis in India is also complicated by stigma, a major barrier to treatment and recovery (Shidhaye & Kermode, 2013).

Given India's severely under-resourced government mental health infrastructure, the country's non-governmental organizations (NGOs) play an important role in delivering mental health care services (Thara & Patel, 2010). These mental health NGOs focus on various mental disorders, such as schizophrenia and substance abuse, and their activities include treatment and rehabilitation, prevention, research, and advocacy (Thara & Patel, 2010). There is scant research, however, on how these organizations use social media platforms to disseminate information about mental health issues, reduce stigma, and engage audiences. This is important to study, given that India has the second-largest number (next to China) of social network users in the world (Ruby, 2023) and also because social media sites are a popular platform worldwide for sharing health information (Moorhead et al., 2013).

In this study, the author analyzes the Instagram posts of three major mental health NGOs in India: White Swan Foundation, Live Love Laugh Foundation, and Mental Health FFoundation India. The author focuses on Instagram because it is a popular image-based mobile application for discussing mental health issues (Muralidhara & Paul, 2018), and India has the largest number of Instagram users in the world at 230.2 million (Statista, n.d.). Further, there is need for more research on how to use images strategically to achieve health outcomes (Vos & Cohen, 2018) because how an illness is portrayed can influence how people with the illness are perceived (Williams, 2012). This study also contributes to our understanding of how non-profits use social media to engage audiences by applying the hierarchy of engagement theoretical framework (Lovejoy & Saxton, 2012), which has been

DOI: 10.4324/9781003328896-9

predominantly applied to non-profits in the United States, to mental health NGOs in the Indian subcontinent. Finally, this study responds to recent calls to integrate a technological affordances perspective into digital public relations scholarship (Zhou & Xu, 2021) by investigating how social media affordances shape Indian mental health NGOs' engagement functions.

To provide context for this study, the author discusses the prevalence and perceptions of mental illness in the Indian context, with particular emphasis on how NGOs deliver mental health care. The author also discusses how these NGOs address stigma by crafting messages that challenge mental illness stereotypes; by harnessing the social media engagement strategies of information, community, and action; and by leveraging social media affordances.

India's Mental Health Landscape

An estimated 60–70 million people in India are affected by mental illness (Ganguli, 2000). Both a biomedical model of mental illness, psychiatry, as well as traditional cultural practices, such as shamanism, mysticism, and indigenous medical systems, such as Ayurveda, are used to understand the cause and treatment of mental illness in India (Kakar, 1982; Wagner et al., 1999). India's mental health infrastructure is severely underdeveloped (Mohandas, 2009). Limited mental health resources are inequitably distributed: Mental health specialists are concentrated in the southern states and in rural areas and the private sector (Thara & Patel, 2010). The country faces a shortage of trained mental health personnel, especially in rural areas, and this lack of resources is attributed to a lack of government funding, inadequate medical training, and negative attitudes toward psychiatrists and psychiatry (Narasipuram & Kasimahanti, 2012; Praharaj et al., 2013).

In the past few decades, NGOs in India have begun to play a role in bridging this gap in mental health care. NGOs can be broadly defined as organizations that are set up with the purpose of providing advocacy and services to the public (Bendana, 2006). The services that mental health NGOs in India provide can be categorized as follows: providing psychosocial rehabilitation, family support, and recreational activities; developing community-based programs and providing services to the destitute and homeless mentally ill; conducting research; and offering training in the form of workshops, seminars, or conferences to health workers, teachers, and students. Further, NGOs advocate for the rights of individuals with mental illness because of the high prevalence of mental illness stigma in India that compromises the rights of these individuals (Ahuja et al., 2017; Thara & Patel, 2010).

Stigma

Stigma is a salient characteristic of India's mental health landscape (Chandra, et al., 2003). Stigma can be broadly defined as an undesirable attribute, visible or invisible, that indicates a spoiled social identity and results in social disapproval (Bos et al., 2013). Stigma can also be defined as a process that occurs when in a situation of power imbalance, individuals are labeled and stereotyped and experience separation, status loss, and discrimination (Link & Phelan, 2013). There are different types of stigma: for example, public stigma refers to people's negative attitudes toward those with a stigmatized condition; self-stigma is when individuals with a stigmatized condition internalize society's negative attitudes toward them; and structural stigma can be defined as stigma perpetuated by institutions and ideologies (Pryor & Reeder, 2011).

People with mental illness are often stigmatized for being dangerousness, unpredictable, incompetent, and for being responsible for their illness (Corrigan & Watson, 2002). In India, mental illness stigma often stems from beliefs that mental illnesses are caused by karma and evil spirits, and there is lack of awareness about biomedical explanatory models of mental illness (Kudva et al., 2020). Such stigma is problematic because it can lead to adverse psychological consequences for the stigmatized person; reduce their access to resources, such as education, employment, and housing; and lead to social isolation (Hatzenbuehler et al., 2013). For example, in the Indian context, individuals with mental illness are hesitant to disclose their illness as it might have an adverse effect on their employment and marriage prospects and may also make them vulnerable to involuntary hospitalization (Kudva et al., 2020). NGOs promote and safeguard the rights of these individuals by disseminating research, lobbying policymakers, and spreading awareness about their work through media such as newsletters, films, and websites (Thara & Patel, 2010).

Media and Mental Illness

The media are a critical factor in influencing attitudes toward mental illness. An analysis of mental illness frames in a major Indian newspaper found that the media perpetuated stigma by associating mental illness with violence, by not capturing the voices of individuals with mental illness, and by the absence of stories on recovery (Subramanian, 2018). However, the stories also challenged stigma by implying socioenvironmental versus individual responsibility for mental illness. Attributing stigmatized conditions, such as mental illness, to external, contextual, socioenvironmental factors rather than internal, individual factors can suggest to audiences that the person with the stigmatized condition is not responsible for their illness and is therefore deserving of empathy (Niederdeppe et al., 2008). This aligns with research that has found that mediated health messages have either a "stigma" or a "challenge" format (Smith, 2007). Messages in a stigma format contain cues such as the person is responsible for their condition, is different from the rest of the community, and is dangerous. On the other hand, messages in a challenge format avoid the generation of stigma by not equating people with the illness but instead promoting social inclusion, hope, and optimism, and encouraging community members "to fight a common problem and to support one another (with or without the health concern) in collective actions addressing testing, research, or treatment" (Smith, 2007, p. 245). Narratives about individuals with stigmatized conditions are a type of challenge message, as they have been found to be effective at inducing compassion toward such individuals by encouraging readers to take the perspective of these individuals, and also show favorable behavioral intentions toward these individuals (e.g., Oliver et al., 2012). The challenge format overlaps with the anti-stigma approaches of education, that is, educating people about mental illness stereotypes/myths and replacing them with factual information; contact, that is, facilitating interpersonal contact with the mentally ill; and protest, that is, raising awareness about various forms of stigma and holding accountable individuals who perpetrate stigma (Corrigan et al., 2012).

Based on the above literature, the author poses the following question:

RQ1: What destigmatizing strategies are reflected in Indian mental health NGO's Instagram communication?

NGOs and Social Media

Social media platforms provide NGOs a low-cost, interactive way to "mobilize supporters, foster dialogic interactions with large audiences, and attract attention to issues that might otherwise be ignored by traditional media" (Guo & Saxton, 2014, p. 60).

This study uses the hierarchy of engagement theoretical framework proposed by Lovejoy and Saxton (2012) and expanded upon by Campbell and Lambright (2020) to investigate how mental health NGOs in India use Instagram to engage with their audiences. According to this framework, social media communication messages by nonprofits have the following three functions: information, community, and action. Informational messages are those that primarily share one-way information about the organization's activities and mission; profile updates and live content (e.g., live tweeting of events) are also classified as informational content. Community messages aim to spark interaction between the organization and its audiences and help the organization create an online community with these individuals. Examples of community messages include the organization thanking its donors or supporters; acknowledging current and local events – for example, holidays; explicitly soliciting responses from followers by posting questions, surveys, contests, and so forth – and replying to comments from the public about them. Action messages are those in which the organization aims to get its audience to perform an action in order to help the organization achieve its financial and strategic goals, and to get audience members to transform from community members to activists and donors. Action messages include those that ask audiences to participate in programming, make a donation, engage in lobbying or advocacy, go to the organization's website, follow it on social media, or repost content.

Research on the application of this framework has found that nonprofits use social media predominantly for communicating information about their work; messages directed toward building community and mobilizing action tend to be fewer (e.g., Huang et al., 2016; Zhou & Pan, 2016). These studies tend to be based on Twitter and Facebook (e.g., Campbell & Lambright, 2020). A recent study of close to 600 US nonprofit organizations' use of social media platforms, such as Instagram, found that these nonprofits use Instagram mainly for disseminating information rather than for community-building or promoting action (Fuller & Rice, 2022). There is scant research on the application of the hierarchy of engagement framework to Indian nonprofits' use of Instagram. Therefore, this study asks:

RQ2: How are the engagement goals of information, community, and action reflected in Indian mental health NGOs' Instagram communication?

This study also investigates how social media affordances shape nonprofit organization's engagement functions by drawing on the theory of technological affordances. First proposed by Gibson (1977), the concept of affordances has been used by communication scholars to describe the characteristics or features of digital media channels that facilitate or constrain human interaction (Zhou & Xu, 2021). Fox and McEwan (2017) classified affordances commonly found in digital media channels as follows: Accessibility refers to a channel's ability to deliver communication regardless of constraints such as time, place, technological literacy (Fox & McEwan, 2017). Bandwidth can be defined as the range of social cues, for example, nonverbal cues, that a media channel can transmit; the structural features of a channel, such as text, photo, audio, or video, enable these cues; because the lack of social cues can lead to misunderstandings, users try to compensate for them, for example, using emoticons in chat rooms (Fox & McEwan, 2017; Zhou & Xu, 2022). Social presence refers to the extent to which a media channel can create the feeling among its users that they are with one another and sharing the same experience (Fox & McEwan, 2017); organizations create social presence by producing humanizing messages (Men et al., 2018). Privacy refers to the extent to which communication on a channel is visible to others; network association is the channel's ability to enable users to identify other users – even if they are geographically distant – through, for example, linked user profiles; personalization is a channel's ability to target a message to particular users; persistence refers to the permanence or durability of content; editability, the extent to which content can be modified before sharing; and conversation control refers to the channel's ability to permit users to manage an online interaction; anonymity or identifiability is the extent to which users feel their true identities can be concealed (Fox & McEwan, 2017).

Based on the above, this study asks the following:

RQ3: How do the technological affordances of social media favor or constrain Indian mental health NGOs' Instagram communication?

Method

Sample

To identify the sample for this study, first, the following search term was entered into the online search engine, google.com, in September 2020: India mental illness mental health organization foundation. The websites of organizations that appeared in the first 10 pages were examined for the following inclusion criteria: national-level organization, focus on mental illness, use of Instagram, at least 100 posts. Exclusion criteria were as follows: a focus on not just India but south-east Asia and/or other countries; a focus on diseases besides mental illness, modeled on a Western mental health organization, only functions as a funding agency.

Of the 11 organizations that the search identified, the following three organizations met these inclusion and exclusion criteria: the White Swan Foundation (White Swan from now onwards), the Live Love Laugh Foundation (Live Love Laugh from now onwards), and the Mental Health Foundation India (Mental Health India from now onwards). For each organization, the time period of sampling was first post until September 30, 2020 (since this was the start date of this study). This resulted in a sample size of 1,098. The three organizations had roughly the same number of Instagram posts: White Swan, 417; Live Love Laugh, 327; Mental Health India, 354. White Swan is a not-for-profit mental health organization founded in 2013 and is focused on mental health awareness through events such as webinars and panel discussions as well as workplace mental health programs (White Swan Foundation, 2021); the organization's partners include India's National Institute of Mental Health and Neurosciences, a major center in the country for education, research, and treatment (National Institute of Mental Health & Neurosciences, 2022). The Live Love Laugh Foundation is a charitable trust founded in 2015 by Indian actor Deepika Padukone; the foundation aims to raise awareness, reduce stigma, and provide resources in the area of mental health (The Live Love Laugh Foundation, 2022; The Live Love Laugh Foundation, 2021). Mental Health India is a not-for-profit organization working in the area of mental health education, training, advocacy, and treatment/care (Mental Health Foundation India, n.d.).

Content and Thematic Analysis

The research questions were answered using content and thematic analysis. The unit of analysis for both the content and thematic analysis was a complete Instagram post, which included both image and caption. As part of pilot coding, a random sample of 90 posts (about 30 from each of the three organizations) was coded to finalize the codebook. This sample was excluded from the final analysis. The remaining posts in the sample (N=1,007) were then coded and then a randomly selected 10 percent of these posts was coded to establish intercoder reliability. Krippendorf's intercoder reliability values were > 0.85. See Table 7.1.

A qualitative thematic analysis was conducted to identify salient patterns of meaning in the data of interest and to develop a story (Neuendorf, 2019). This process involved using a constant-comparative method to identify initial codes that reflected key features of the data and collapsing codes into themes (Glaser & Strauss, 1967; Neuendorf, 2019). The process was guided by the following sensitizing concepts identified in previous research on mental illness and health stigma (Gupta & Ariefdjohan, 2021; Smith, 2007) and on the critical media effects framework (Ramasubramanian & Banjo, 2020): responsibility attribution, inclusionary/

Table 7.1 Operational Definitions of Content Analysis Variables

Variable	*Definition*
Language	Consider language in text, image, caption, and hashtags 1-English/Mainly English 2-Mainly language other than English 3-Both English and other languages
Hashtags	Total number of unique hashtags in the entire post, i.e., both image and text
Comments	Total number of audience comments
Public Replies	Total number of public replies by the organization
User Mentions	Total number of unique user mentions in the entire post, i.e., both image and text
Likes	Total number of likes

Source: Table created by the author.

exclusionary language, and power/voice. The thematic analysis also drew upon the concepts of information, community, and action from the hierarchy of engagement framework (Lovejoy & Saxton, 2012) and affordances theory (Fox & McEwan, 2017) to explore how nonprofits communicatively use Instagram to destigmatize mental illness.

Results

RQ1: What destigmatizing strategies are reflected in Indian mental health NGOs' Instagram communication?

Posts by both Live Love Laugh and White Swan reflected a primary aim to destigmatize mental illness by providing education/information in a didactic format in the captions of Instagram posts about a range of mental health issues affecting various groups of people. For example, in Live Love Laugh, a series called #AtoZofDepression discussed how depression is different from anxiety, how bullying in adolescence can lead to depression, and how meditation can relieve stress and anxiety. Other topics covered by this organization included myths and facts on self-injury and self-harm; suicide; signs of mental health issues in children; stress, anxiety, and depression in the elderly; post-partum depression; children, parents', and teachers' mental health during the pandemic; workplace mental health; and depression in the LGBTQ community.

Similarly, White Swan covered a range of mental health issues such as caregiver mental health, substance abuse, mental health in the workplace/workplace well-being, mental health and romantic relationships, mental health and grief, borderline personality disorder, mental health and exam stress, mental health and COVID-19, mental health in the elderly, and suicide.

A socioenvironmental attribution for the cause and treatment of mental health issues was noticeable in the posts of the Live Love Laugh and White Swan NGOs.

Through a series of posts marked with the hashtag #NotJustHormones, the White Swan NGO aimed to provide information about the psychosocial causes of mental health issues in women. For example, a White Swan post had two images with the following text:

> What people think postpartum depression is caused by – hormones. What postpartum depression is actually caused by: hormones, lack of proper nutrition to the mother, societal pressures, preexisting mental health struggles, multitasking, lack of access to mental health professionals, high stress environment, sleeplessness and exhaustion, diet restrictions, and lack of support from family members or partner.

Another series of posts by the organization on workplace mental health points to the role of employers in safeguarding and nourishing their employees' mental health. For example, in one post, a healthcare professional explains the

> RESPECT model, a framework to support employee mental health: regular contact, empathetic communication, supportive communication, practical help, encourage help seeking from a mental healthcare provider, consider providing support for someone who has taken a mental health break, tell them that help is available.

White Swan also destigmatized seeking help for mental health issues by drawing parallels between mental health and physical health issues. For example, a series of posts demystified visiting mental healthcare professionals. The caption of one post said: "We know exactly what to expect, what information to go with and how to prepare for a visit to a medical professional for a physical illness but we know very little about visiting a psychiatrist." The image accompanying the caption provided questions psychiatrists might pose on the first visit, such as, "What are your symptoms and how long have you had them? Do they happen occasionally or all the time? Do you have a history of illnesses or a family history?" Content in White Swan resonated with the voice and image of a range of clinical experts who were typically identified by their name and title/qualifications, for example, "Neurospsychologist Tanvi Mallya," "Dr. Aishwarya is a clinical psychologist in private practice," and "Aarti Jagannathan, assistant professor, department of psychiatric social work, NIMHANS" [National Institute of Mental Health and Neurosciences]. Posts about treatment, however, tended to focus on psychosocial interventions like therapy. In-depth information about psychiatric medications was rarely, if ever, provided.

While White Swan destigmatized mental illness by humanizing mental health professionals, Live Love Laugh provided a range of empathetic first-person narratives of mental illness. These narratives showcased a diverse range of mental health issues (e.g., bipolar disorder, paranoid schizophrenia, clinical depression, personality disorder, suicide, PTSD, body-image issues, obsessive compulsive disorder, and relationship issues). By giving audiences insight into the lived experiences of individuals with mental illness, they concretized the often abstract and intangible symptoms of mental illness and put a human face on such individuals. In these narratives, the protagonist typically explained the causes, symptoms, diagnosis, and treatment of their disorder. For example:

> I was 24, when I diagnosed with Major Depression. I just lost my interest in each and everything. It started getting worse when i was facing the fears of getting contaminated, being so obsessed with orders and symmetry and trypophobia. My thoughts were not in my control and I was feeling groggy and the thoughts of self killing occupied my mind!

In another narrative, an individual with mental illness explained what their treatment looked like:

> The treatment was simple. He would give me some pills that I had to take in the morning and in the night. It would take a few weeks to kick in, but after that along with therapy once a week, I would feel a difference he assured me. … The pills made me tired, but it was helping. My heart felt lighter.

Overall, the personal narratives were optimistic in tone, emphasized empathy for those with mental illness, and indicated that living with a mental illness is possible.

RQ2: How are the engagement goals of information, community, and action reflected in Indian mental health NGOs' Instagram communication?

Posts by both Live Love Laugh and White Swan reflected a primary engagement goal of providing information to destigmatize mental illness, as discussed above. For the Live Love Laugh and White Swan NGOs, a secondary communicative function of Instagram posts appeared to be mobilizing audiences to raise awareness and destigmatize mental illness.. For example, one post read:

> How often do you hear your colleagues say these phrases at work? Do you want to help and don't know how? We're launching a unique program called #WorkplaceMentalHealthChampions to help you talk about mental health at work. This program will give you the skills you need to facilitate conversations around mental health at work. Sign up now! Link in the bio.

Another post encouraged the audience to be allies to individuals with mental health issues:

> Together, we can create a community that is stigma-free, discrimination-free, supportive of people with mental health issues. If we all take a pledge to be a responsible ally, 85 percent Indians who do not seek help for mental health issues will be able to reach out. Will you take the pledge to be one of the

> #OctoberAllies? If yes, visit the link in our bio and take the pledge. You can download the image and post it on your Instagram. Tag your friends to ask them to take the pledge as well. Let's build a community of mental health allies!

The NGO Mental Health Foundation appeared to use its Instagram account primarily as a marketing and promotional platform to advertise and encourage attendance at its various mental-health-related events: a "Mental Health Festival," workshops, panel discussions, and colloquiums. Regarding the festival, posts advertised a range of festival activities such as a short film competition, a slam poetry event, a crafts competition, a live concert, and a puppetry show. Posts provided information about festival early-bird discounts and other incentives and festival registration information as well as festival sponsors. For example:

> Mental Health Foundation (India) brings to you the three most awaited events of India's Biggest Mental Health Festival – Storytelling, Hindi Poetry & Poetry Slam! This is a chance for you to engage and experience the magic that words can have at our Curated Open Mic on themes of Mental Health at AIIMS, New Delhi. Discounted Registrations close today, so hurry now and register yourself to see the magic unfurl!

Other events advertised by the organization on its Instagram included a youth summit on mental health; a mental health colloquium; community mental health camps; a mental healthcare online program; and panel discussion events with topics such as how mental health issues are ignored in the education system, the mental health of caregivers, and violence in interpersonal relationships.

Although across all three NGOs, audiences interacted with the organization via comments, the NGOs typically did not reply to the commenters, even when the audience posed a direct question to the organization (See Table 7.2). For example, this comment on a Live Love Laugh post did not receive a reply: "@tlllfoundation I suppose I have Bipolar disorder. I checked out the symptoms and I've got at least fourteen of 'em. How can I get in touch with you?" This is surprising, given that the inclusive language used by both White Swan and Live Love Laugh NGOs in their posts (e.g., "EveryoneACaregiver" and "YouAreNotAlone") indicated the desire

Table 7.2 Audience Engagement Variables

Organization	*Mean Hashtags (M, SD)*	*Mean Audience comments (M, SD)*	*Public Replies (M, SD)*	*Likes (Median)*	*Unique User Mentions (M, SD)*
Live Love Laugh	5.23, 4.45	11.83, 8.57	1.11, 2.18	1297	0.68, 1.34
Mental Health India	4.71, 4.11	1.21, 2.56	.06, 0.40	85	0.98, 1.67
White Swan	6.43, 5.10	0.49, 0.94	.04, 0.22	25	0.74, 2.38

Source: Table created by the author.

Table 7.3 Language Used in Posts

Organization	*Mainly English (%)*	*Mainly language other than English (%)*	*Both English and other languages (%)*	*Total Posts*
Live Love Laugh	296 (97.69%)	6 (1.98%)	1 (0.33%)	303
Mental Health India	309 (96.87%)	7 (2.19%)	3 (0.94 %)	319
White Swan	378 (98.18%)	3 (0.78%)	4 (1.04%)	385

Source: Table created by the author.

to foster a feeling of community and a collective commitment to educate about mental illness and reduce stigma. Other strategies to create a sense of online community were also not predominant. For example, Mental Health Foundation India attempted to foster a sense of community by acknowledging current events like "World Suicide Prevention Day," but such posts were not common. Finally, online community building was negatively affected by the fact that across all three NGOs, the posts were predominantly in English (See Table 7.3). This limited audience interaction as only audiences who are conversant in this language can comprehend the NGOs' posts and interact with them.

RQ3: How do the technological affordances of social media favor or constrain Indian mental health NGOs' Instagram communication?

Both Live Love Laugh and White Swan used the Instagram affordances of text, image, and video to communicate and educate about mental health issues. For example, White Swan normalized mental health care by helping audiences visualize mental health treatment through video interviews with psychiatrists and psychologists and also through photographs of these experts. Live Love Laugh used the platform's affordances of captions, image, and video to provide a range of empathetic first-person narratives of mental illness. These digital storytelling practices helped both organizations cultivate their social presence.

Both the Live Love Laugh and White Swan NGOs used hashtags as digital activist tools to raise awareness of their goals, coalesce discussion around them, and motivate action. For example, the White Swan Foundation used the hashtag "OctoberAllies" to encourage audiences to pledge to become mental health activists. The NGO used the "#WorkplaceMentalHealthChampions" hashtag to promote a program.

Prominent hashtags used by the Live Love Laugh NGO included "TogetherAgainstDepression" and "#YouAreNotAlone." Another prominent hashtag was "#NotAshamed." Instagram posts tagged with this hashtag provided visual portraits of everyday professionals with a particular mental illness. For example, "Rindamma, housewife, 54 years, mental illness survivor" and "Sandhya, journalist, 39 years, mental illness survivor." The posts aimed to raise awareness that mental illnesses affect people of various backgrounds, and the #notashamed hashtag aimed to reduce stigma and to encourage people to openly share their mental health story: “Childhood trauma can manifest into depression in late years. Watch Anovshka’s story to learn more. #NotAshamed. Visit the link mentioned in our bio to know more and join the movement.”

Both Live Love Laugh and White Swan also engaged in cross-platform promotion of their work. For example, “Log on to www.thelivelovelaughfoundation.org to find out more about anxiety and its relationship with depression.” Like the Live Love Laugh and White Swan NGOs, Mental Health Foundation also used hashtags to raise awareness about its events, for example, #MentalHealthFestival and #MHFI. See Figures 7.1, 7.2, and 7.3 for word clouds of hashtag frequencies for each NGO.

As mentioned earlier, the NGOs typically did not reply to audience comments, even when the audience posed a direct question to the organization (See Table 7.2). This underutilization of Instagram’s affordance of interactivity translated to a noticeable absence of a consistent dialogic loop in the posts, wherein the organization solicited feedback and answered questions from audiences. Indeed, when the White Swan Foundation attempted to interact with the audience by posting a question (“What do you think contribute to women being more vulnerable to disorders such as depression and anxiety?”), the audience did not respond with any comments, reflecting the organization’s failure to establish a social presence through interactive dialogue. Finally, Instagram’s affordances of interactivity and accessibility were limited by the fact that posts across all three NGOs were predominantly in English (See Table 7.3). In a

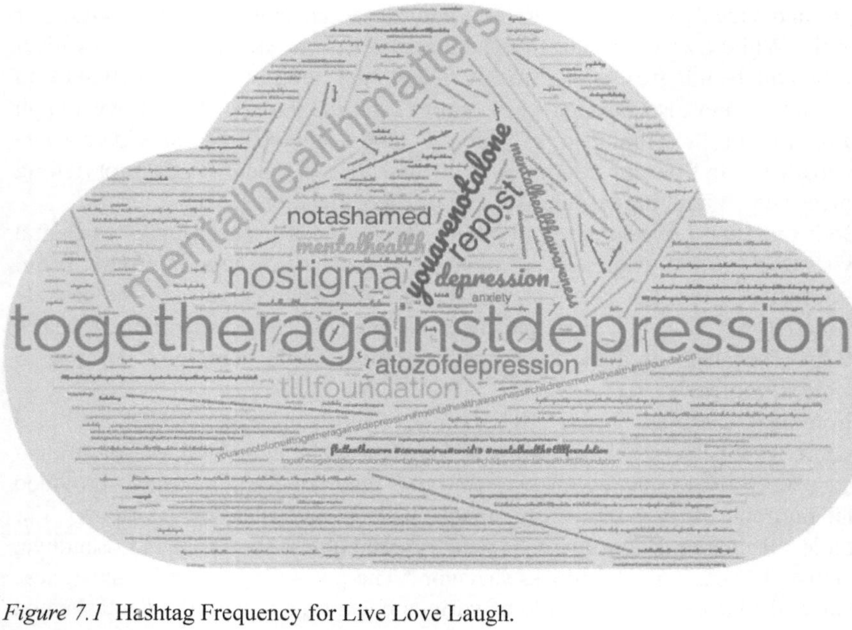

Figure 7.1 Hashtag Frequency for Live Love Laugh.

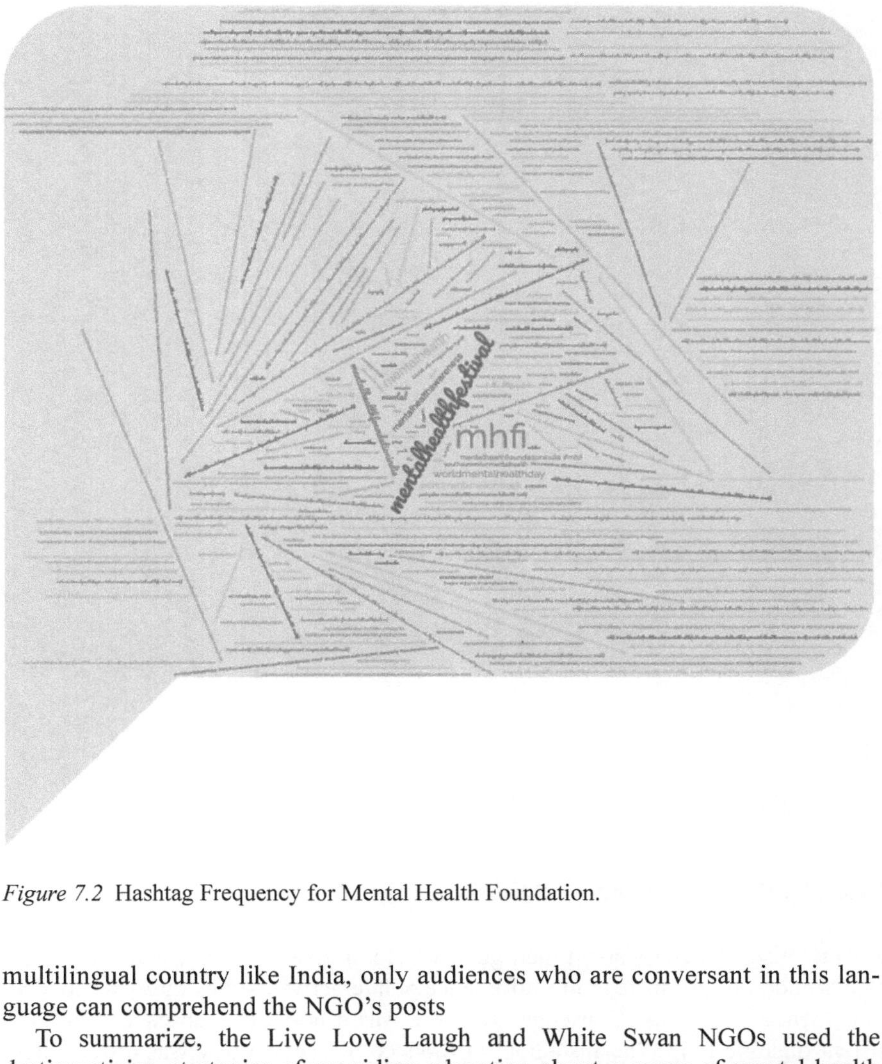

Figure 7.2 Hashtag Frequency for Mental Health Foundation.

multilingual country like India, only audiences who are conversant in this language can comprehend the NGO's posts

To summarize, the Live Love Laugh and White Swan NGOs used the destigmatizing strategies of providing education about a range of mental health issues as well as providing mediated contact with both mental healthcare professionals and individuals with lived experiences of mental illness via humanizing visual narratives. In terms of engagement functions, Live Love Laugh and White Swan used social media primarily to provide information, whereas Mental Health India used Instagram primarily as a promotion and marketing tool to mobilize audiences. All three NGOs disengaged from online community-building. In terms of Instagram affordances, both Live Love Laugh and White Swan used text, image, and video to engage in digital storytelling about mental health and cultivate a social presence. Further, all three NGOs used hashtags as digital activist

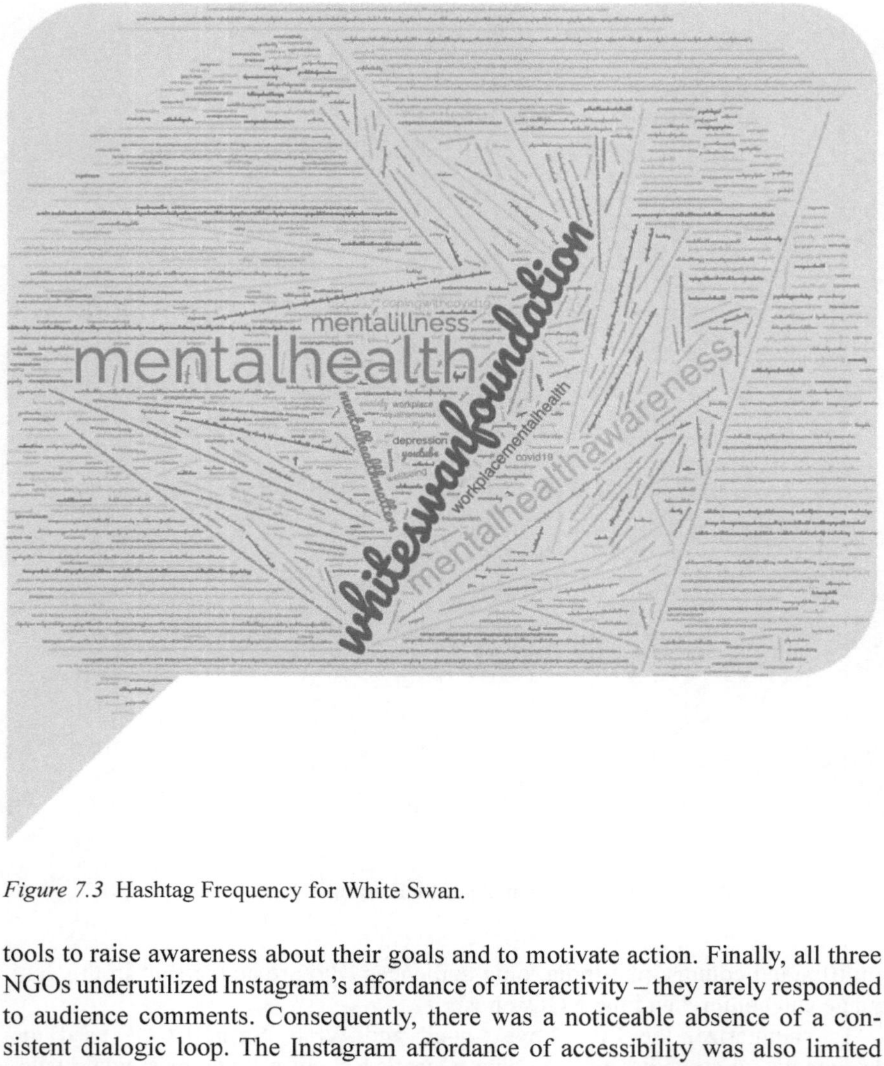

Figure 7.3 Hashtag Frequency for White Swan.

tools to raise awareness about their goals and to motivate action. Finally, all three NGOs underutilized Instagram's affordance of interactivity – they rarely responded to audience comments. Consequently, there was a noticeable absence of a consistent dialogic loop. The Instagram affordance of accessibility was also limited by the fact that, across all three NGOs, posts were mainly in English, even though India is a multilingual country.

Discussion

Drawing on mental illness stigma research and literature on nonprofit organizations' engagement strategies and social media affordances, this study investigated, via a content and thematic analysis, how three major mental health NGOs in India communicatively use Instagram to destigmatize mental illness.

Similar to previous research on nonprofits engagement strategies, this study found that both White Swan and the Live Love Laugh NGOs primarily used Instagram to provide information to audiences about mental illness. The informational posts by White Swan and Live Love Laugh NGOs were about a range of mental health issues affecting various groups of people. Such posts about a diverse range of topics can help discourage a monolithic conceptualization of mental illness and make audiences aware of the various manifestations of this group of disorders. The posts also reflected these NGOs' intersectional approach to addressing mental illness as they grappled with how individuals' marginalized social identities such as gender, sexual orientation, and age, can exacerbate their mental health issues.

The posts tended to emphasize socioenvironmental causes and treatments for mental illnesses and, in doing so, highlighted the prevalence of structural stigma and the need for structural-level interventions. Emphasizing the social determinants of mental health can reduce stigma by shifting attributions of responsibility – for both the cause and solution of a health problem – away from an individual and toward society (Fisher & Baum, 2010). While White Swan informational posts tended to foreground the voice of mental health professionals, Live Love Laugh centered the lived experiences of people with mental illness in its informational posts. Narratives on mental illness have been found to reduce stigma by helping audiences empathize with the individuals in the story (e.g., Oliver et al., 2012). Both NGOs used the Instagram affordances of caption, image, and video to humanize information.

For both White Swan and Live Love Laugh NGOs, a secondary function of Instagram posts was to mobilize audiences to engage in collective efforts to combat stigma. Both organizations used hashtags to make their content easily searchable and to coalesce audiences around specific topics (Guo & Saxton, 2014).

The Mental Health Foundation primarily used Instagram to mobilize audiences to attend its events. All three NGOs tended to not utilize Instagram for online community building; specifically, they rarely engaged in interactive dialogue with audiences. This last finding aligns with previous research on Western non-profit organizations, which has found that these organizations tend to not exploit the community-building function of social media (Lovejoy & Saxton, 2012).

It is noteworthy that, despite the multilingual landscape of the country, all three organizations use predominantly English in their Instagram posts. Given that English is the language of the educated middle class in India (Rubdy, 2018), this finding corroborates criticism that NGOs in India serve middle-class populations and do not address grassroots issues (Brown, 2014).

Theoretical and Practical Implications

This study contributes to the limited literature on nonprofits' social media communication in a non-Western/non-Eurocentric context. The study expands the hierarchy of engagement framework by examining it through the lens of the destigmatizing strategies of education, contact, and protest (Corrigan et al., 2012). In doing so, the study adds nuance to the hierarchy of engagement framework by identifying additional information types that mental health NGOs share with their audiences – specifically, information about the socioenvironmental/structural causes and treatments for mental illness and narratives about the lived experiences of people with mental illness. This finding has practical value as information about structural causes and interventions can reduce stigma by deflecting blame for mental illness from individuals to society. Further, providing mental health information in the form of first-person multimedia narratives can reduce stigma by making concrete the abstract and intangible symptoms of mental illness and by humanizing individuals with mental illness. Other practical implications include NGOs increasing the use of multiple languages in their messaging to audiences in order to expand their reach; using hashtags to increase their social presence; and responding to audience comments to build an online community.

Limitations and Future Research

This study is limited to an analysis of three non-profit mental health NGOs in India. In future studies, it would be worthwhile to expand the scope of this research with a larger sample size. In future research, it would also be worthwhile to investigate how the different communicative functions of these NGOs and their destigmatizing strategies relate to audience engagement variables such as comments and likes. It would also be helpful to investigate what challenges NGOs encounter in using social media platforms to build online and offline relationships with multilingual publics.

Conclusion

This exploratory study provides a preliminarily step to understanding how mental health NGOs in India use Instagram to destigmatize mental illness. Key findings include that the NGOs engage audiences mainly by providing information about mental health issues on Instagram and by mobilizing them to combat stigma. The study expands nonprofits' hierarchy of engagement framework by identifying socioenvironmental causes and treatments for mental illness and narrative content as additional information types. The NGOs tended to not use Instagram for community building; there was lack of a dialogic loop with audiences, and the mainly English-language content of the posts limited audience accessibility in a multilingual country. The NGOs used the Instagram affordance of hashtags to raise the visibility of their content and as a digital activist tool.

Acknowledgments

The author would like to thank Julia Quigley, MA, for assistance with coding.

References

Ahuja, K. K., Dhillon, M., Juneja, A., & Sharma, B. (2017). Breaking barriers: An education and contact intervention to reduce mental illness stigma among Indian college students. *Psychosocial Intervention*, *26*(2), 103–109.

Bendaña, A. (2006). NGOs and social movements (A North/South divide?). *Civil Society and Social Movements*, 22. Retrieved from www.rrojasdatabank.info/Bendana.pdf

Bos, A. E., Pryor, J. B., Reeder, G. D., & Stutterheim, S. E. (2013). Stigma: Advances in theory and research. *Basic and Applied Social Psychology*, *35*(1), 1–9.

Brown, T. (2014). Negotiating the NGO/social movement dichotomy: Evidence from Punjab, India. *International Journal of Voluntary and Nonprofit Organizations*, *25*(1), 46–66.

Campbell, D. A., & Lambright, K. T. (2020). Terms of engagement: Facebook and Twitter use among nonprofit human service organizations. *Nonprofit Management and Leadership*, *30*(4), 545–568.

Chandra, P. S., Deepthivarma, S., Carey, M. P., Carey, K. B., & Shalinianant, M. P. (2003). A cry from the darkness: Women with severe mental illness in India reveal their experiences with sexual coercion. *Psychiatry: Interpersonal and Biological Processes, 66,* 323–334. doi:10.1521/psyc.66.4.323.25446

Corrigan, P. W., & Watson, A. C. (2002). Understanding the impact of stigma on people with mental illness. *World Psychiatry*, *1*(1), 16.

Corrigan, P. W., Morris, S. B., Michaels, P. J., Rafacz, J. D., & Rüsch, N. (2012). Challenging the public stigma of mental illness: a meta-analysis of outcome studies. *Psychiatric Services*, *63*(10), 963–973.

Dandona, R., Kumar, G. A., Dhaliwal, R. S., Naghavi, M., Vos, T., Shukla, D. K., Vijayakumar, L., Gururaj, G., Thakur, J. S., Ambekar, A., Sagar, R., Arora, M., Bhardwaj, D., Chakma, J. K., Dutta, E., Furtado, M., Glenn, S., Hawley, C., Johnson, S. C., … Dandona, L. (2018). Gender differentials and state variations in suicide deaths in India: The Global Burden of Disease Study 1990–2016. *The Lancet Public Health*, *3*(10), e478–e489. https://doi.org/10.1016/S2468-2667(18)30138-5

Fisher, M., & Baum, F. (2010). The social determinants of mental health: implications for research and health promotion. *Australian & New Zealand Journal of Psychiatry*, *44*(12), 1057–1063.

Fox, J., & McEwan, B. (2017). Distinguishing technologies for social interaction: The perceived social affordances of communication channels scale. *Communication Monographs*, *84*(3), 298–318.

Fuller, R. P., & Rice, R. E. (2022). Nonprofit organization communication, crisis planning, and strategic responses to the COVID-19 pandemic. *Journal of Philanthropy and Marketing*, *27*(3), e1750.

Ganguli, H. C. (2000). Epidemiological findings on prevalence of mental disorders in India. *Indian Journal of Psychiatry, 42,* 14–20.

Gibson, J. J. (1977). The Theory of Affordances. *Hillsdale, NJ*, *1*(2), 67–82.

Glasser, B. G., & Strauss, A. L. (1967). *The Discovery of Grounded Theory: Strategies for qualitative research.* Aldine Transaction Publishers.

Guo, C., & Saxton, G. D. (2014). Tweeting social change: How social media are changing nonprofit advocacy. *Nonprofit and Voluntary Sector Quarterly*, *43*(1), 57–79. https://doi.org/10.1177/0899764012471585

Gupta, R., & Ariefdjohan, M. (2021). Mental illness on Instagram: A mixed method study to characterize public content, sentiments, and trends of antidepressant use. *Journal of Mental Health*, *30*(4), 518–525.

Hatzenbuehler, M. L., Phelan, J. C., & Link, B. G. (2013). Stigma as a fundamental cause of population health inequalities. *American Journal of Public Health*, *103*(5), 813–821.

Huang, Y., Lin, Y., & Saxton, G. (2016). Give me a like: How HIV/AIDS nonprofit organizations can engage their audience on Facebook. *AIDS Education and Prevention*, *28*, 539–556.

Kakar, S. (1982). *Shamans, mystics and doctors: A psychological inquiry into India and its healing traditions*. University of Chicago Press.

Kudva, K. G. El Hayek, S., Gupta, A. K., Kurokawa, S., Bangshan, L., Armas-Villavicencio, M. V. C., .. & Sartorius, N. (2020). Stigma in mental illness: Perspective from eight Asian nations. *Asia-Pacific Psychiatry*, *12*(2), e12380.

Link, B. G., & Phelan, J. C. (2013). Labeling and stigma. In *Handbook of the sociology of mental health* (pp. 525–541). Springer.

Lovejoy, K., & Saxton, G. D. (2012). Information, community, and action: How nonprofit organizations use social media. *Journal of Computer-Mediated Communication*, *17*(3), 337–353

Mohandas, E (2009). Roadmap to Indian psychiatry. *Indian Journal of Psychiatry, 51*, 173–179. doi:10.4103/0019-5545.55083

Moorhead, S. A., Hazlett, D. E., Harrison, L., Carroll, J. K., Irwin, A., & Hoving, C. (2013). A new dimension of health care: Systematic review of the uses, benefits, and limitations of social media for health communication. *Journal of Medical Internet Research*, *15*(4), e85. https://doi.org/10.2196/jmir.1933

Muralidhara, S., & Paul, M. J. (2018). Healthy selfies: Exploration of health topics on Instagram. *JMIR Public Health & Surveillance*, *4*(2), e10150. https://doi.org/10.2196/10150

Narasipuram, S., & Kasimahanti, S. (2012). Quality of life and perception of burden among caregivers of persons with mental illness. *Andhra Pradesh Journal of Psychological Medicine, 13*, 99–103.

National Institute of Mental Health & Neurosciences. (2022). *About Us*. https://nimhans.co.in/about-us/

Neuendorf, K. A. (2019). Content analysis and thematic analysis. In P. Brough (Ed.), *Research Methods for Applied Psychologists: Design, analysis and reporting* (pp. 211–223). Routledge.

Niederdeppe, J., Bu, Q. L., Borah, P., Kindig, D. A., & Robert, S. A. (2008). Message design strategies to raise public awareness of social determinants of health and population health disparities. *The Milbank Quarterly*, *86*(3), 481–513.

Oliver, M. B., Dillard, J. P., Bae, K., & Tamul, D. J. (2012). The effect of narrative news format on empathy for stigmatized groups. *Journalism & Mass Communication Quarterly*, *89*(2), 205–224.

Praharaj, S. K., Behere, R. V., Deora, S., & Sharma, P. S. V. N. (2013). Psychiatric specialization as an option for medical students in the Indian context. *International Review of Psychiatry, 25*, 419–424. doi:10.3109/09540261.2013.822347

Pryor, J. B., & Reeder, G. D. (2011). HIV-related stigma . In B. Hall, J. Hall, & C. Cockerell (Eds.), *The AIDS Epidemic: Treatment and epidemiology in the 21st century* (pp. 790–803). Shelton, CT: PMPH-USA.

Ramasubramanian, S., & Banjo, O. O. (2020). Critical media effects framework: Bridging critical cultural communication and media effects through power, intersectionality, context, and agency. *Journal of Communication*, *70*(3), 379–400.

Rubdy, R. (2018). Code alternation and entextualization in bilingual advertising: The construction of glocal identities in India's Amul Butter ads. In *Language and Literature in a Glocal World* (pp. 29–56). Springer.

Sagar, R., Dandona, R., Gurugaj, G., Dhaliwal, R. S., Singh, A., Ferrari, A., Dua, T., Ganguli, A., Varghese, M., Chakma, J. K., Kumar, G. A., Shaji, K. S., Ambekar, A., Rangaswamy, T., Vijayakumar, L., Agarwal, V., Krishnankutty, R. P.,Bhatia, R., Charlson, F., … Dandona, L. (2019). The burden of mental disorders across the states of India: The global burden if disease study 1990–2017. *The Lancet*, *7*(2), 148–161. https://doi.org/10.1016/S2215-0366(19)30475-4

Sharma, S., Kundu, A., Basu, S., Shetti, N. P., & Aminabhavi, T. M. (2020). Indians vs. COVID-19: The scenario of mental health. *Sensors International*, *1*, 100038. https://doi.org/10.1016/j.sintl.2020.100038

Shidhaye, R., & Kermode, M. (2013). Stigma and discrimination as a barrier to mental health service utilization in India. *International Health*, *5*(1), 6–8. https://doi.org/10.1093/inthealth/ihs011

Smith, R. (2007). Medical depictions of health topics: Challenge and stigma formats. *Journal of Health Communication*, *12*(3), 233–249. https://doi.org/10.1080/10810730701266273

Subramanian, R. (2018). Frames of mental illness in an Indian daily newspaper. *Health Communication*, *34*(14), 1806–1815. https://doi.org/10.1080/10410236.2018.1536948

Thara, R., & Patel, V. (2010). Role of non-governmental organizations in mental health in India. *Indian Journal of Psychiatry*, *52*(1), 389–395. https://doi.org/10.4103/0019-5545.69276

The Live Love Laugh Foundation. (2021, December 6). *LiveLoveLaugh study points to significant shifts in India's perception of mental health* [Press Release]. Retrieved from https://dwdja6ta65an1.cloudfront.net/the_live_love_laugh/uploads/media/source/CicDCA-LiveLoveLaugh-Study-Points-to-Significant-Shifts-in-Indias-Perception-of-Mental-Health.pdf

The Live Love Laugh Foundation. (2022). *About the Foundation*. Retrieved from www.thelivelovelaughfoundation.org/about-us

Vos, S. C., & Cohen, E. L. (2018). Using pictures in health and risk messages. In R. Parrott (Ed.), *The Oxford Encyclopedia of Health and Risk Message Design and Processing*. Oxford University Press. Retrieved from https://oxfordre.com/communication/display/10.1093/acrefore/9780190228613.001.0001/acrefore-9780190228613-e-314;jsessionid=A40869530585B72F550B81C834A2C283

Wagner, W., Duveen, G., Themel, M., & Verma, J. (1999). The modernization of tradition: Thinking about madness in Patna, India. *Culture & Psychology*, *5*, 413–445. doi:10.1177/1354067X9954003

White Swan Foundation. (2021). *About White Swan Foundation*. Retrieved from www.whiteswanfoundation.org/about-us

Williams, I. (2012). Graphic medicine: How comics are revolutionizing the representation of illness. *Hektoen International: A Journal of Medical Humanities*, *3*(4). www.graphicmedicine.org/online-articles/graphic-medicine-how-comics-are-revolutionizing-the-representation-of-illness/

Zhou, A., & Xu, S. (2021). Digital public relations through the lens of affordances: a conceptual expansion of the dialogic principles. *Journal of Public Relations Research*, *33*(6), 445–463.

Zhou, H., & Pan, Q. (2016). Information, community, and action on Sina-Weibo: How Chinese philanthropic NGOs use social media. *Voluntas: International Journal of Voluntary and Nonprofit Organizations*, *27*, 2433–2457.

Part III

Mobile Communication in the Public Sphere

8 How Mobile Users Differ from Non-Mobile Users in #IndiaFightsCorona on Twitter

An LDA Topic Modeling and Content Analysis

Md Enamul Kabir and Louisa Ha

Introduction

Mobile communication has replaced many dyadic face-to-face communications in our daily lives. Beginning with voice and short message service (SMS), mobile communication now dominates the vast majority of information and media outlets, bringing a revolution to the modes of communication accessibility and frequency (Schrock, 2015). Most importantly, it changed our way of voicing opinions and supporting public issues. Due to the high penetration of mobile devices and apps, the use of social media platforms has expanded significantly over the past few years.

Because people refrained from physical contact during the Covid-19 outbreak, activism and public health awareness campaigns have predominantly shifted to social media platforms such as Twitter, Facebook, and Instagram. Capitalizing on this trend, smartphones, serving as portable Internet communication devices, took the opportunity to provide users with engaging, addictive, and unparalleled experiences. These enhanced user experiences extended beyond entertainment and convenience. User experiences and user behaviors between mobile devices and laptops/desktops differ due to portability, accessibility, and speed of typing on these devices (Palin et al., 2019). These characteristics have a significant impact on the communication experiences of social media users. In fact, Twitter reports that due to the surge in mobile phone usage, mobile users are 46 percent more inclined to generate unique tweets compared to users of other devices (Twitter Marketing, 2013). However, Twitter research did not study the difference in the kind of postings made by mobile and desktop/laptop users.

Additionally, mobile users may exhibit specific personality features themselves. Past studies such as Shaw et al. (2016) have shown that mobile users of the two most popular smartphone platforms, Android and iPhone iOS, exhibited distinct personality traits online. Android users are more inclined to be modest, whereas iPhone users are more extroverted, more adventurous, more attention-craving, more educated, and are comparatively wealthier as the cost of iPhones is much higher than most Android phones (Hixon, 2014; Sanjeev et al., 2019). Studies

DOI: 10.4324/9781003328896-11

also indicate that social media users' personality qualities influence their online activity (Brailovskaia & Bierhoff, 2020). Chen and Caropreso (2004) showed that online communication style, pattern, and task involvement are influenced by user personality. Such patterns can be analyzed to uncover how various mobile phone users differ from non-mobile users in their communication style on Twitter (Hixon, 2014; Shaw et al., 2016).

The social media platform Twitter is an effective resource for health promotion campaigns (Wehner et al., 2014). The American Cancer Society, the American Heart Association, and the American Diabetes Association are just a few of the many health groups that utilize Twitter to raise money and awareness for their respective causes (Park et al., 2016). In fact, Twitter's core platform was built from the ground up to work well on mobile devices. Intending to cater to mobile users, Twitter has developed a condensed text version and implemented various accessibility options on its platform (Smith & Brenner, 2012). Despite the prevalence of mobile use in modern social media, researchers have rarely delved deeper into the comparison of mobile and non-mobile sources on Twitter.

Recent research has shed light on the efforts of Indian public health promotion regarding Covid-19 through Twitter, particularly focusing on the discourse involving the hashtag #IndiaFightsCorona (Shah & Sebastian, 2020). In light of the rising COVID-19 cases in India, the government took steps to both implement humanitarian and economic measures and promote awareness of their efforts. Starting from March 2020, the hashtag #IndiaFightsCorona was utilized by the Indian Ministry of Information and Broadcasting (MIB) as a means to effectively communicate and share information regarding their public health efforts during the pandemic (Shah & Sebastian, 2020). The hashtag #IndiaFightsCorona was already trending on Twitter, and it continued to sustain since the ministry account was created. Indians have used the hashtag to express their opinion about the promotion of Indian health and the prevention of disease. The variety of users' communication patterns, particularly those observed during the COVID-19 pandemic, can provide insight into how technological affordance influences online health communication. Consequently, it is reasonable to expect a variation in communication style between the health-promoting Tweets of Indian Android phones and iPhone users, especially since iPhone users are much more affluent than Android phone users. The goal of this chapter is to investigate how mobile users (e.g., Android and iPhone) differ from non-mobile users (webapp) on Twitter regarding the use of various smartphone operating system (OS) users in Indian health promotion to prevent COVID-19 infection.

Theoretical Framework

Technology Affordance of Mobile Communication

The concept "mobile technology affordance" pertains to the possibilities and constraints of the everyday usage of mobile devices. James Gibson (1977) pioneered the concept of "affordance," which was an integral aspect of his ecological psychology theory. Gibson referred to the interaction between the material world and the human being, suggesting that we are designed to work in harmony with our natural surroundings. Based on Gibson's idea, it is possible to engage in a wide variety of actions in any given location (Raudaskoski, 2003).

Extending the scope of affordances theory, Schrock (2015) proposed the concept of "communicative affordances," with a communication-focused approach, as "an interaction between subjective perceptions of utility and objective qualities of the technology that alter communicative practices or habits" (p. 1232). A clearer picture of affordance in light of mobile communication can be found in Hutchby's (2001) definition of communicative affordance, which "frames the practices through which technologies come to be involved in the weave of ordinary conduct" (p. 450). To determine the affordances that are communicated, Gaver (1991) suggested that the technological artifact can be analyzed to see how they function following such attributes. According to Schrock (2015), how mobile media transforms patterns of communication today is better understood with technology affordance theory than with previous technology-related theories. This theory identified four mobile media affordances: portability, availability, locatability, and multimediality. The portability factor in particular explains why smartphone popularity has surpassed that of non-mobile electronic devices. Now that smartphones have almost the same processing power as desktop computers, the key distinction between mobile media and desktop media is portability because of the small size of mobile media. Next, the availability factor determines how users maintain their relationships with different people and how they tend to communicate with their closest friends via multiple channels, including texting, voice calls, and social media (Schrock, 2015). These factors make the affordance theory extremely relevant to the various facets of mobile communication on Twitter. Rathnayake and Suthers (2017) extended the discussion further, explaining that Twitter hashtags can be viewed as an availability affordance for two reasons. First, hashtags can be created and retrieved, and second, diverse outcomes result from their use, including discussion, expression, measurement of engagement, observation of activity, and so forth. Other features, capabilities, and evolutions (such as the "Like" button, and retweet option) on social media platforms that are specifically designed for mobile phones (e.g., Twitter), may open up new avenues for behaviors (Bucher & Helmond, 2018). Past Research also indicated that the users' cultural identities and levels of personal experience, and the social setting in which they live all play a role in establishing their affordances (Gaver, 1991; Hutchby, 2001). This research investigates the socio-economic and linguistic factors that determine the Twitter use of users with various smartphone operating systems (such as Android and iPhone) in connection to health-promotion efforts.

Information Flow Theory (IF)

Information flow theory is relatively new to the field of social media research, but it has been effectively employed in understanding Twitter engagement (Hemsley, 2019). Defining the idea of virality as the process through which a piece of information, such as a tweet, gets disseminated across a network of individuals far removed from its original source, Nahon and Hemsley (2013) conceptualized the framework of IF theory as the convergence of common interests between average users and a select group of influential users who control the dissemination of information. In this flow, the influence of a user is evaluated with the idea that those with the most views, shares, or retweets often reach the greatest number of people. This is similar to the opinion leader theory (two-step flow theory), in which some individuals serve as gatekeepers and influencers of other people. The media is a conduit via which influential people learn about issues, and these people in turn inform the population at large. This indicates that most people get their information from influential people through personal connections rather than through the media at large (Liu, 2007). It promotes the idea that some users in our social networks have more influence over the flow of information than others through a greater number of followers and retweets.

Mobile OS Platforms

A mobile operating system (OS) is the basic software necessary to operate smartphones, tablets, and other mobile computing devices. The development of operating systems has progressed from PC-based OS to embedded OS to the most cutting-edge operating system based on smartphones. This evolutionary growth in mobile OS is primarily caused by technological advancements in hardware, software, and the Internet (Okediran et al., 2014). In this study, Apple's iOS and Google's Android platforms were explored, since these are the two most popular mobile operating systems in the current smartphone market (Kohli et al., 2020). Additionally, these smartphone OS platforms became relevant since the company started identifying whether a user tweets from a desktop or mobile device and, if mobile, whether they used Twitter's iOS or Android apps or a third-party service in their log data. Researchers can get more context while interpreting the sources/devices where the tweets are posted. It is important to note that Twitter was established targeting mobile phone users limiting the text to only 140 characters. Since smartphones reached human hands, our online engagement has increased tremendously. Social media platforms have grown to become storming sources of activism. According to Twitter, thousands of hashtags are being generated and posted per hour every day (Twitter Marketing, 2013). Such enormous engagement on Twitter was observed only after the smartphones like Android and iPhone were available to people of various socio-economic groups.

Socio-Economic Factors Behind Varying Mobile OS

The Android smartphone includes a range of devices from high-end "over a thousand" dollars to low-cost "under a hundred" devices. As a result, Android users come from a wide range of socio-economic backgrounds. In contrast, most people who use Apple iOS are probably better off financially because of the higher cost of iPhones. But it is hard to determine the social and economic status of people who use webapps from any non-mobile devices or personal computers. Yet those who use computers are typically more educated and are white-collar professionals. Also typing from computers allows thoughtful thinking and writing with the comfort of the keyboard. Categorizing the behaviors of people's mobile communication into these groups based on their OS use allows insights about their socio-economic groups as well. The socio-economic difference between the two groups is consistent with the observation that 58 percent of smartphone sales in India are in the $150 category (below 10,000 rupees). Whereas, due to the price-sensitive nature of the Indian market, iOS-based smartphones account for just 3 percent of the Indian smartphone market (Sanjeev et al., 2019).

The massive dominance of Android and iOS (mobile) users over web app (computer/non-mobile) users in social media engagement nowadays provides a wealth of information. The Twitter participation of mobile users has risen sufficiently to become significant, and they, therefore, merit scholarly consideration. However, the engagement behavior of Android and iOS phone users on Twitter remains unstudied. It is expected that Android users would dominate Twitter since India's mobile operating system market is dominated by Android, with a 95.26 percent share (Statista, 2023b). In contrast, it would be interesting to see how the affluent iPhone users in Indian society contribute to the Tweets, as they only account for 3.1 percent of the population. These classifications can be used to examine how engaged these different socio-economic groups are with health promotion tweets from India. In other words, it can offer insights such as which operating system users/socio-economic classes are more interested in tweets about health.

According to Jiang and Ha (2020), each information-carrying gadget is a conduit that helps individuals gather, organize, and disseminate data in health promotion. Due to their unique characteristics and user experiences, several electronic gadgets might be considered alternative channels for disseminating information. Besides, although identical information may be seen on both desktop computers and smartphones, the two devices elicit very distinct feelings and reactions from their users (Cho et al., 2014) because computers have bigger displays, and smartphones are more convenient and simpler to use than computers. To that end, it stands to reason that users' perceived convenience, affordance, and socio-economic background would determine what platform they would prefer for specific topics.

Twitter for Health Promotion in India

In India, mobile communication has become an increasingly important medium. India is among Asia's top countries for mobile phone penetration, right behind China (Statista, 2022). Furthermore, the enormous smartphone industry has made it feasible for all families in India in a variety of socio-economic statuses to purchase smartphones with a wide price range. With 1.39 billion users, India's smartphone industry is one of the biggest in the world, and 90 percent of those phones use the Android operating system (Statista, 2017).

According to Statista (2022), India has risen to the number three spot in the world in terms of Twitter usage. In light of evidence suggesting that Android, iPhone iOS users, and webapp users have different online personas (Hixon, 2014; Sanjeev et al., 2019; Shaw et al., 2016), it is necessary to examine whether and to what extent smartphone users' Twitter communication habits differ from each other. At this juncture, exploring the nature of webapp (laptop or desktop computers), Android, and iPhone usage of Indian users to express their opinions concerning public health promotion combating COVID-19 would uncover the impact of their technology affordance on health communication on social media platforms. The technology affordance theory offers insights into the various devices users choose for tweeting, but the effectiveness of those OS platforms can be understood through the conceptual framework of Information flow (IF) theory (Raudaskoski, 2003; Nahon & Hemsley, 2013). As IF theory suggested, users who have the most followers and number of retweets are assumed to have the most influence on other users (Hemsey, 2019). In addition, various research showed the interconnectedness and centrality as shown by the number of followers and retweets as an indicator of user influence on Twitter (Freelon & Karpf, 2015; Hemsley, 2019). According to Hemsley (2016), retweets typically attract more followers, which can increase the reach of subsequent tweets. Among the various mobile OS platforms (i.e., Android, iOS) and non-mobile devices used by Indian users, the following research questions probe which platform users have more influence in terms of retweets in the #IndiaFlightsCorona health campaign.

RQ1: Which platform (iOS/Android/webapp) users receive more retweets in the hashtag #IndiaFightsCorona?
RQ2: Which platform (iOS/Android/webapp) users have a higher number of followers in the hashtag #IndiaFightsCorona?

The purpose of this study is to compare the Twitter use of different smartphone platform users (e.g., Android and iPhone) and webapp users (laptop/desktop) in Indian health promotion in combatting Covid-19. The tweets containing the hashtag #IndiaFightsCorona are an excellent resource for obtaining an in-depth investigation of the phenomena. Especially, it would be intriguing to explore the underlying themes in the discussions comparing users of smartphone OS platforms (e.g., iOS and Android). This gives way to the following research question,

RQ3: Is there a significant difference in the expressed topics between Android users and iPhone users in the Tweets associated with the hashtag ##IndiaFightsCorona?

Language Complexity in India

Among the millions of tweets in India every day, 50 percent are usually posted in English (Mandavia & Krishnan, 2019). The reason is that English is the most common second language and the native language used by Twitter and other common platforms. Hence, English tweets have a global reach and are targeted at a global audience. In contrast, tweets posted in local (vernacular) language are intended to target a local audience. Past media research in India has typically focused on English content. However, the language issue is quite unique in India's smartphone communication. First of all, despite India having over 22 official languages throughout the country, none of them are considered the national language (Pool, 1991). English is the very widely spoken language in India and is prevalent in the realms of business, education, and government documentation. Indian people also put a high value on English for their children's education and the path to career success (Mukherjee & Bernaisch, 2020). Indeed, India is considered the second-largest English-speaking nation in the world after the United States. Yet the India Human Development Survey (2005) indicated that only around 28 percent of Indian people spoke at least some English (Desai et al., 2010).

Second, the central government initiated the hashtag #IndiaFightsCorona, targeting the whole Indian population. Thus, more hashtags must be observed in the language that is generally used in government operations. For the same reason, the English tweets associated with this hashtag might have been retweeted more than non-English tweets. Due to the elitist nature of the English language in India and probably a global audience of English tweets, we propose this:

H1: English tweets will have a higher number of retweets than non-English Tweets.
H2: English tweeters will have a higher number of followers than non-English Tweeters.

In addition, it has not been examined among the various smartphone OS users (e.g., iOS and Android) which group tends to use English more. This study offers insights into the language preferences of these OS-based user groups. Undeniably, this preference is varyingly determined by the socio-economic and educational background of these users. In general, iOS users are more educated than Android users (Hixon, 2014). A recent study conducted by the online advertising network Chitika also found that states with more college graduates tend to have higher iPhone sales. There is no published study about the language preference of various smartphone OS users in India. However, studies indicated that, in India, people who are wealthier, more educated, and upper caste were found to be more likely to use English, as the Lok Foundation survey showed (Rukmini, 2019). Indian men who are fluent in English earn, on average, 34 percent more in hourly wages compared to men who do not speak English. Men who speak some English also earn 13 percent more in hourly wages compared to those who do not speak English (Azam et al., 2013). A substantial difference may likely be observed in the preferred languages used by users of Android and iOS (Sanjeev et al., 2019). The following hypothesis is proposed,

H3: English Tweeters will be more likely to use iOS smartphones than non-English Tweeters.

Research Design and Method

Data Collection

The data collection was performed with a social media data extractor and analyzer app Netlytic from Twitter (Gruzd, 2016). Data were collected from May 2022 through September 2022. The app Netlytic, initially developed by Anatoliy Gruzd at Ryerson University (now Toronto Metropolitan University), is now widely used in human behavior research (Santarossa, 2019; Meneses, 2019). In addition, the tweets with the hashtag #IndiaFightsCorona were collected using the Twitter application programming interface (API), which allows researchers to extract source data of the tweet such as Twitter for Android, Twitter for iPhone, and Twitter webapp (non-mobile devices).

In all, 40,000 tweets with the hashtag #IndiaFightsCorona were gathered. After cleaning duplicate tweets and processing the text, 32,000 tweets remained in the dataset. Due to the size of the dataset, Python programming language was used to code and label the data. This technology, unlike manual content analysis, allows for the examination of a bigger dataset.

Measurement

The variables are measured as follows – source platform/device (android/iOS/webapp, etc.), language, number of followers, and number of retweets. The following are the variables measured for this study.

Source platform. Twitter offers information about the sources of the tweets. However, it comes in an unstructured manner. When inquired using the Python programming command for unique information about tweet sources, over 50 categories turned up among which most were bots. All the tweets were coded into four categories, tweets from the Android platform were coded 0, tweets from the iOS platform were coded as 1, tweets from non-mobile devices were coded as 2, and finally, all the tweets from Twitter bots were put together into one category and coded as 3.

Language. Two language types were coded. With Python's uniqueness inquiry command, all the languages were explored in the 32,064 tweets. Tweets that used English were coded as 1, and the tweets that used other languages were coded as 0.

Data Analysis

Unsupervised machine learning was used to test the hypotheses and answer the research questions regarding the topic of the tweets. In particular, Latent Dirichlet Allocation (LDA) topic modeling was employed to answer RQ3.

LDA Topic Modeling

Latent Dirichlet allocation (LDA) topic modeling is a computational content analysis method for uncovering the hidden thematic structure of a corpus of texts (Maier et al., 2018). It uses the Variational Exception Maximization (VEM) approach to derive the greatest likelihood estimate from the entire corpus of the text. For the analysis, LDA models use a probabilistic process that is both hypothetical and abstract, necessitating a range of assumptions topic modeling as latent structures in a text corpus have been shown to be an effective method for rapidly identifying large thematic clusters in large text corpora (Guo et al., 2016). An advantage of using a topic model over simple co-occurrence analysis is that it can unearth a hidden semantic relationship between words even if they have never appeared in the same document. LDA topic modeling's mixed membership approach is an additional benefit when compared to other topic-modeling methods (Maier et al., 2018).

Guo et al. (2016) expressed concern about employing unsupervised learning in Twitter data because "tweets are constrained to be short pieces of text (no longer than 140 characters) and are often terse, truncated, and quite 'messy' because they contain abbreviations, symbols, and intentionally truncated words, in addition to spelling errors and poor grammar" (p. 337). In that case, compared to other models, one advantage LDA has is that it is flexible enough to allow aggregating the entire Twitter archive into a single document, compiling from a certain period, or a combination of both (Ha et al., 2022). This technique makes the analysis easier in the LDA topic model. Moreover, recent research has critiqued those traditional models such as dictionary-based analysis for their lack of depth, and being time-consuming, and labor-intensive (Onan, 2022). In LDA topic models, the distribution of themes characterizes a document, and the distribution of words describes a topic. This allows for a far more lucid understanding of the interdependencies across the topics.

Preprocessing and Lemmatization

The dataset was preprocessed and lemmatized using the NLTK toolkit in the Python computer language. For an algorithm to interpret any language, the text must be broken down into smaller pieces that our computer can comprehend (Kabir, 2022). In natural language processing, stemming and tokenization serve this goal. First, stemming was used to remove suffixes (such as "ing," "ly," "es," and "s") from a word. For instance, "play," "player," "played," "plays," and "playing" are the several forms of the word "play." As part of the preprocessing and cleaning, the entire dataset was stemmed to turn the words to their base form (e.g., pen, pens, pens', pen's→ pen). In addition to that, as Guo et al. (2016) suggested, all the punctuations, numbers, extra spaces, special characters, and stop words were also removed.

Table 8.1 Device Use in #IndiaFightsCorona Tweets

Source Platform	*Frequency*	*Percent*	*Cumulative Percent*
Tweets from Android	19248	60	60
Tweets from iPhone	1679	5.2	65.3
Tweets from Webapp	7426	23.2	88.4
Tweets from Bots	3711	11.6	100
Total	32064	100	

Source: Table created by the authors during analysis.

Table 8.2 Comparison of Retweets between Users of Different Devices

Retweets-descriptive

	N	*Mean*	*Std. Dev.*	*Std. Err*	*95% Confidence Interval*		*Min*	*Max*
					Lower Bound	*Upper Bound*		
Android	19248	456.11	3076.37	22.18	412.65	499.58	0	31767
iOS	1679	163.74	1386.90	33.85	97.35	230.12	0	31767
webapp	7426	142.28	1709.53	19.83	103.40	181.17	0	31767
Twitter bots	3711	4.31	29.53	0.49	3.36	5.26	0	350
total	32064	315.83	2547.53	14.22	287.95	343.72	0	31767
Model Fixed effects			2547.51	12.20	288.01	343.65		
Random effects				152.70	-170.15	801.81		

Source: Table created by the authors during analysis.

Results

The descriptive analysis showed that, among the 32,064 tweets, a massive 19,248 came from Android-based smartphones (60%), 1,679 tweets came from iOS (5.2%), 7,426 came from webapp (23.2%), and 3,711 were posted by Twitter bots (11.6%) (See Table 8.1).

To answer research question RQ1, a one-way between-subjects ANOVA was conducted to compare the difference between smartphone OS platforms and the number of retweets for iOS, Android, webapp, and bots. There was a significant effect of smartphone OS platforms on the number of retweets at the $p<.001$ level for the four conditions [$F(3, 32060) = 51.68$, $p < .001$] (See Table 8.2). Android users received the highest number of retweets (Mean = 456). iPhone users and computer users have similar mean numbers of retweets.

To answer the RQ2, a one-way between-subjects ANOVA was conducted to compare the effect of smartphone OS platforms on the number of followers for iOS, Android, webapp, and bots (see Table 8.3). There was a significant effect of smartphone OS platforms on the number of followers at the $p<.001$ level for the four conditions [$F(3, 32060) = 298.727$, $p < .001$] (Table 8.2). Computer users who tweet on the topic have a much higher number of followers ($M = 175{,}585$) than iPhone ($M = 46{,}185$) and Android users ($M = 14{,}549$). Most remarkably, Twitter bots also have a large number of followers ($M = 125{,}435$).

Next, we examined whether English tweets rendered a higher number of retweets than non-English Tweets, and whether non-English tweeters had a lower number of followers than English Tweeters as posited in H1 and H2 respectively (see Table 8.4). The t-test result shows that English tweets rendered a significantly higher number of retweets ($M = 412.19$, $SD = 3109.72$) than non-English Tweets ($M=136.88$, $SD= 730.846$), $t(32062) = 9.242$, $p < .001$) Thus, the hypothesis H1 that English tweets render a higher number of retweets was supported.

The t-test result shows that English tweeters rendered a significantly higher number of followers ($M = 7{,}962{,}906$, $SD = 415702.782$) than non-English Tweeters ($M=416{,}44.98$, $SD = 427{,}156.165$), $t(32062) = 7.729$, $p < .001$ (see Table 8.5). Thus, hypothesis H2 was also supported.

To test the H3, a chi-square test of independence was performed to examine the relation between the device and the language of tweets. The relation between these variables was significant, $c^2 = 1101.911$, $p < .0001$ (See Table 8.6). English Tweeters (68%) were more likely to use iPhone users than non-English Tweeters (32%).

Finally, to answer RQ3, LDA topic modeling was employed to determine whether tweets by iOS users featured topics different from those found in tweets from Android users. Several parameters, such as the number of topics to be detected, were adjusted to build an ideal topic model that reveals interpretable patterns in each topic's words. The LDA topic model revealed six topics from each OS group of users. Based on the words appearing in each topic, labels were assigned to define the underlying themes (see Table 8.7). Some of the common themes that emerged from the analysis of the tweets from both groups were the information about COVID-19 vaccines and the vaccination progress, appreciation of the health workers, advice about safety measures, and so forth These were also the most dominant topics with more or less similar keywords. However, some unique topics emerged from the two user groups in the analysis. Tweets from iOS contained two unique topics such as concerns about COVID-19-linked deaths and vaccination updates (which are of interest to the broader population outside of India). In contrast, the two unique topics that emerged from the tweets of Android users were in reference to publicity of the Indian government and local politics.

There was also a noticeable semantic difference between iOS and Android users in a few tweets. Comparatively, to express their appreciation, Android users used phrases like "completed" and "touched" that evoke emotion and sympathy to convey their appreciation of health professionals, while iOS users used words like "largest vaccination drive" and "warriors."

Table 8.3 Comparison of Device Users and Number of Followers

User_followers Descriptives								
	N	*Mean*	*Std. Dev.*	*Std. Err*	*95% Confidence Interval*		*Min*	*Max*
					Lower Bound	*Upper Bound*		
Android	19248	14548.71	181497.40	1308.21	11984.51	17112.92	0	5341102
iOS	1679	46185.55	386997.08	9444.58	27661.17	64709.94	0	7053718
webapp	7426	175585.47	586759.33	6808.99	162237.93	188933.02	0	18843324
Twitter bots	3711	125435.69	745857.94	12243.64	101430.76	149440.62	0	18148034
total	32064	66335.10	420131.04	2346.26	61736.35	70933.86	0	18843324

Source: Table created by the authors during analysis.

Table 8.4 Retweets Comparison between English and Non-English Tweets

	Lang	*N*	*Mean*	*Std. Deviation*	*Std. Error Mean*
Retweet count	English	20842	412.18	3109.72	21.55
	Non-English	11222	136.89	730.85	6.90

Source: Table created by the authors during analysis.

Table 8.5 User Follower Comparison between English and Non-English Tweets

	lang	*N*	*Mean*	*Std. Deviation*	*Std. Error Mean*
user_followers_count	English	20842	79629.06	415702.78	2879.48
	Non-English	11222	41644.98	427156.17	4032.29

Source: Table created by the authors during analysis.

Table 8.6 Language and Device Comparison

			android	*iOS*	*webapp*	*twitter bots*	*Total*
language	Non-English	Count	7897	537	2271	517	11222
		% within source	41.0%	32.0%	30.6%	13.9%	35.0%
	English	Count	11351	1142	5155	3194	20842
		% within source	59.0%	68.0%	69.4%	86.1%	65.0%
Total		Count	19248	1679	7426	3711	32064
		% within source	100.0%	100.0%	100.0%	100.0%	100.0%

Source: Table created by the authors during analysis.

Discussion

The findings from this study indicate that, among the smartphone operating systems (OS), Android users have higher retweets than iOS users in Indian Covid-19 combatting promotion through the hashtag #IndiaFightsCorona. However, non-mobile users have the highest number of followers, followed by Twitter bots. This shows that influencers are more likely to be professionals who type their tweets carefully on a computer rather than a mobile phone.

The findings also indicate that the language in the tweets was a significant indicator of the number of retweets and followers and, in turn, the degree of influence.

Table 8.7 Topic Models of Tweets from iOS and Android

Tweets from iOS	*Tweets from Android*
Covid19 vaccine Information	Covid19 vaccine Information
[far, recovery, rate, covid19, active, doses, administered, india, currently]	[caseload, doses, recovery, stands, active, administered, rate, far, currently]
Appreciation of health workers	Appreciation of health workers
[largestvaccinationdrive, personnel, amp, warriors, airport, cisf, we4vaccine, unite2fightcorona, covid19, bhubaneswar]	[touched, completed, people, million, hardwork, know, coronawarriors, harghartiranga, amp]
Vaccination progress	Vaccination progress
[vaccination, administered, vaccines, group, age, years, cr, india]	[1st, dose, age, administered, vaccines, years, cr, india]
Vaccination Update	Publicity for Government
[update, pm, india, covid, doses, uts, states, crore, vaccine]	[covid, doses, country, largestvaccinedrive, vaccine, narendramodi, covid19, crore, pm, india]
Covid-19 concern	Local politics
[start, deaths, claims, believing, regarding, lohittezusunpura, today, india, covid]	[number, june, new, telangana, radheshyam, delhi, andhrapolitics, newsfeed, indiapolitics, breakingnews]
Safety Advice	Safety Advice
[covid19, precaution, safe, pandemic, appropriate, stay, mask, unite2fightcorona, largestvaccinedrive, covid]	[wear, unite2fightcorona, safe, behavior, stay, appropriate, mask, pandemic, covid]

Source: Table created by the authors during analysis.

More specifically, we found English tweets were more influential than non-English tweets regarding the number of retweets and followers. The fact that the number of English tweets almost doubled non-English tweets shows the importance of English tweets. This is consistent with previous studies and links education and language in relation to iOS users (Hixon, 2014).

The lexical differences found in the topic model between iOS and Android users are also worth noting. The topic "Publicity for the Government" came from Android users and indicated keywords referring to the conservative government in India. A positive sentiment toward the government and the current conservative prime minister was observed while tracing the keywords back in the Tweets. For example, one Tweet under the topic "publicity" wrote *"Under the dynamic leadership of our Hon,Äôble PM Shri @narendramodi Ji, India,Äôs Cumulative COVID-19 Vaccination Coverage exceeds 196.32 Cr now. #IndiaFightsCorona."* Although, there is not enough evidence to show the political inclination of Indian Android/iPhone/webapp users (i.e., conservative, progressive/liberal), expressions of Android users who support conservative politicians offer insights and room for future studies. Research conducted in the United States by Localytics found that 70 percent of the states with the most iPhone users vote for the Democratic party, while 70 percent of the states with the most Android phone users vote for the Republican party (Roe, 2012). Though Localytics' study was based on the United States, the current government in India is also conservative and demonstrated deep ties with the conservative government during Trump's presidency. Besides, *Social Times*' research showed that people who use Android devices are 20 percent more inclined to adhere to conservative values than iPhone users (Hoffmire, 2017). This leaves room for investigation for future research regarding what degree iOS or Android users might be inclined to various political values across the world. These discoveries provide practical implications for health researchers as well. For instance, given the observed variation in user behavior between smartphone operating systems and webapp users, social media analytics researchers and educators of the future must be mindful of this variation while analyzing Twitter health promotional data. Especially considering the substantial differences in political inclinations and socio-economic standings between Android and IOS users. In addition, the language preference of iOS users can help healthcare professionals better comprehend the outreach of health promotion campaigns in countries such as India, where both English and non-English are widely utilized on social media.

Conclusion

The purpose of this study was to examine the Twitter use of different smartphone platform users (e.g., Android and iPhone) and webapp users (laptop/desktop) in Indian health promotion in combatting COVID-19. The tweets containing the hashtag #IndiaFightsCorona were collected and analyzed using both statistical and an unsupervised machine learning model (LDA topic model) to obtain in-depth insight into the phenomena. Some of the topics the LDA topic model revealed were referencing COVID-19 vaccines and the vaccination progress, appreciation of the health workers, advice about safety measures, and so forth. In addition, this study contributes to more understanding of English versus non-English Indian Twitter users. While presenting various ways smartphone operating systems influence the Twitter sphere, this study is not without its limits. First, motivation or intent to use an operating system (e.g., iOS and Android) could be crucial to this study. However, it was not possible to reach out to each Twitter user and interview or survey their motivation behind the use of each mobile platform. Thus, that motivation factor is missing from this research. However, future research can accomplish this by communicating a survey questionnaire to a smaller sample of Twitter users regarding their smartphone operating system preferences. A project of this nature can provide answers to research questions pertaining to user motivation for mobile technology use and elucidate the technological affordances of South Asian mobile users. Second, topic modeling helps classify huge amounts of textual data, but it does not capture nuances that might be investigated via manual content analysis. Although this study encapsulates various angles and perspectives of iOS and Android users' tweets, including language, and socio-economic factors, a separate qualitative inquiry would reveal deeper insights. Thus, this study recommends a manual thematic analysis for future research. It is also important to note that the relationship between platform use and behavior is not causal. That is, the differences in mobile platforms use such as iOS and Android do not cause the differences in the aforementioned behavior. Rather, it just shows that people more active on Twitter were found to select certain operating systems. However, the use of computer typing by influencers shows that they are more cautious in writing their tweets than those smartphone users.

References

Ajzen, I. (1991). The theory of planned behavior. *Organizational Behavior and Human Decision Processes, 50*(2), 179–211.

Azam, M., Chin, A., & Prakash, N. (2013). The returns to English-language skills in India. *Economic Development and Cultural Change, 61*(2), 335–367.

Brailovskaia, J., & Bierhoff, H. W. (2020). The narcissistic millennial generation: A study of personality traits and online behavior on Facebook. *Journal of Adult Development, 27*(1), 23–35.

Broniatowski, D. A., Jamison, A. M., Qi, S., AlKulaib, L., Chen, T., Benton, A., ... & Dredze, M. (2018). Weaponized health communication: Twitter bots and Russian trolls amplify the vaccine debate. *American Journal of Public Health, 108*(10), 1378–1384.

Bucher, T., Helmond, A. (2018). The affordances of social media platforms. In Burgess J., Marwick A., Poell T. (Ed.), *The SAGE Handbook of Social Media* (pp. 233–253). SAGE.

Chen, S. J., & Caropreso, E. J. (2004). Influence of personality on online discussion. *Journal of Interactive Online Learning, 3*(2), 1–17.

Cho, W., Jung, Y., & Im, J. H. (2014). Students' evaluation of learning management systems in the personal computer and smartphone computing environments. *International Journal of Mobile Communications, 12*, 142–159.

Desai, S. B., Dubey, A., Joshi, B. L., Sen, M., Shariff, A., & Vanneman, R. (2010). *Human development in India*. Oxford University.

Dixon, S. (2022, July 27). *Twitter: number of monthly active users 2010–2019*. Retrieved from www.statista.com/statistics/282087/number-of-monthly-active-twitter-users/

Dunn, A. G., Surian, D., Dalmazzo, J., Rezazadegan, D., Steffens, M., Dyda, A., Leask, J., Coiera, E., Dey, A. & Mandl, K. D. (2020). Limited role of bots in spreading vaccine-critical information among active twitter users in the United States: 2017–2019. *American Journal of Public Health, 110*(S3), S319-S325.

Dyer, O. (2018). Vaccine safety: Russian bots and trolls stoked online debate, research finds. *BMJ: British Medical Journal (Online), 362*. www.bmj.com/content/362/bmj.k3739.full

Freelon, D., & Karpf, D. (2015). Of big birds and bayonets: Hybrid Twitter interactivity in the 2012 presidential debates. *Information, Communication & Society, 18*(4), 390–406.

Gaver, W. W. (1991, March). Technology affordances. In *Proceedings of the SIGCHI conference on human factors in computing systems* (pp. 79–84).

Gibson, J. J. (1977). *The Theory of Affordances*. Erlbaum Associates.

Gruzd, A. (2009). Studying collaborative learning using name networks. *Journal of Education for Library and Information Science, 50*(4), 243–253.

Gruzd, A. (2016). Netlytic: Software for automated text and social network analysis. Available at http://Netlytic.org.

Guo, L., Vargo, C. J., Pan, Z., Ding, W., & Ishwar, P. (2016). Big social data analytics in journalism and mass communication: Comparing dictionary-based text analysis and unsupervised topic modeling. *Journalism and Mass Communication Quarterly, 93*(2), 398–415. https://doi.org/10.1177/0264550519880595

Ha, L. Ray, R., Matanji, F., & Yang, Y. (2022). How news media content and fake new about the trade war are shared on twitter: Topic modeling and content analysis. In Ha, L. & Willnat, L. (Eds.) *The US-China Trade War: Global news framing and public opinion in the digital age* (pp. 125–144). Michigan State University Press.

Hampton, K. A. (2010). Internet use and the concentration of disadvantage: Glocalization and the urban underclass, *American Behavioral Scientist, 53*(8), 1111–1132. http://abs.sagepub.com/content/53/8/1111

Hemsley, J. (2016). Studying the viral growth of a connective action network using information event signatures. *First Monday, 21*(8). https://doi.org.ezproxy.bgsu.edu/10.5210/fm.v21i8

Hemsley, J. (2019). Followers retweet! The influence of middle-level gatekeepers on the spread of political information on twitter. *Policy and Internet, 11*(3), 280–304. https://doi.org/10.1002/poi3.202

Hixon, T. (2014, April 10). What kind of person prefers an iPhone? Retrieved February 21, 2019, from www.forbes.com: www.forbes.com/sites/toddhixon/2014/04/10/what-kind-of-person-prefers-an-iphone/#57282023d1b0

Hoffmire, J. (2017, Jan 3). What your smartphone says about who you are. Retrieved from www.deseret.com/2017/1/3/20603300/john-hoffmire-what-your-smartphone-says-about-who-you-are.

Hutchby, I. (2001). Technologies, texts and affordances. *Sociology, 35*(2), 441–456.

Jane, M., Hagger, M., & Foster, J. (2018). Social media for health promotion and weight management: a critical debate. *BMC Public Health, 18*, 932 .https://doi.org/10.1186/s12889-018-5837-3

Jiang, W., & Ha, L. (2020). Smartphones or computers for online sex education? A contraception information seeking model for Chinese college students. *Sex Education, 20*(4), 457–476. https://doi.org/10.1080/14681811.2019.1672041

Kabir, M. (2022). Topic and sentiment analysis of responses to Muslim clerics' misinformation correction about COVID-19 vaccine: Comparison of three machine learning models. *Online Media and Global Communication*. https://doi.org/10.1515/omgc-2022-0042

Kahlor, L. (2010). PRISM: A planned risk information seeking model. *Health Communication 25*(4): 345–356.

Katz, E., Blumler, J. G., & Gurevitch, M. (1974). Utilization of mass communication by the individual. In J. G. Blumler & E. Katz (Eds.), *The Uses of Mass Communications: Current perspectives on gratifications research* (pp. 19e32). SAGE.

Kohli, H. S., Khandai, S., & Gulla, A. (2020). Smartphone operating system preference based on different personality & lifestyle traits of the consumer. *International Journal of Scientific & Technology Research, 9*(2), 2444–2449.

Liu, F. C. (2007). Constrained opinion leader influence in an electoral campaign season: Revisiting the two-step flow theory with multi-agent simulation. *Advances in Complex Systems, 10*(02), 233–250.

Maier, D., Waldherr, A., Miltner, P., Wiedemann, G., Niekler, A., Keinert, A., Pfetsch, B., Heyer, G., Reber, U., Häussler, T., Schmid-Petri, H., & Adam, S. (2018). Applying LDA topic modeling in communication research: Toward a valid and reliable methodology. *Communication Methods and Measures, 12*(2–3), 93–118. https://doi.org/10.1080/19312458.2018.1430754

Mandavia, M. & Krishnan, R. (2019, Nov 11). "Non-English tweets are now 50% of the total": Twitter India MD. Retrieved from https://economictimes.indiatimes.com/industry/tech/non-english-tweets-are-now-50-of-the-total-twitter-india-md/articleshow/72000048.cms

Meneses, L. (2019). Netlytic. *Early Modern Digital Review, 2*(1). https://doi.org/10.7202/1064533ar

Mukherjee, J., & Bernaisch, T. (2020). The development of the English language in India. In Kirkpatrick, A. *The Routledge Handbook of World English* (pp. 165–177). Routledge.

Nahon, K. & J. Hemsley. (2013). *Going Viral*. Polity.

Norman, D. A. (2008, November 17). *Affordances and Design*. https://jnd.org/affordances_and_design/

Okediran, O. O., Arulogun, O. T., Ganiyu, R. A., & Oyeleye, C. A. (2014). Mobile operating systems and application development platforms: A survey. *International Journal of Advanced Networking and Applications, 6(*1), 2195.

Onan, A. (2022). Bidirectional convolutional recurrent neural network architecture with group-wise enhancement mechanism for text sentiment classification. *Journal of King Saud University-Computer and Information Sciences, 34*(5), 2098–2117.

Palin, K., Feit, A. M., Kim, S., Kristensson, P. O. & Oulasvirta, A. (2019, October). How do people type on mobile devices? Observations from a study with 37,000 volunteers. In *Proceedings of the 21st International Conference on Human-Computer Interaction with Mobile Devices and Services* (pp. 1–12).

Park, H., Reber, B. H., & Chon, M. G. (2016). Tweeting as health communication: health organizations' use of Twitter for health promotion and public engagement. *Journal of Health Communication, 21*(2), 188–198.

Pool, J. (1991). The official language problem. *American Political Science Review, 85*(2), 495–514. doi:10.2307/1963171

Rathnayake, C., & Suthers, D. D. (2017, July). Twitter issue response hashtags as affordances for momentary connectedness. In *Proceedings of the 8th International Conference on Social Media & Society* (pp. 1–10).

Raudaskoski, S. (2003). The affordances of mobile application. In *Workshop on Technology, Interaction and Workplace Studies* (pp. 08–09).

Roe, M. (2012, Oct 3). Infographic: Who do iPhone users vote for? Android users? *KPCC.* Retrieved from https://archive.kpcc.org/blogs/politics/2012/10/03/10295/iphone-states-vote-democrat-android-states-vote-re/

Rukmini, S. (2019, May 14). In India, who speaks in English, and where? *Mint.* Retrieved from www.livemint.com/news/india/in-india-who-speaks-in-english-and-where-1557814101428.html

Sanjeev, M. A., Sehrawat, A., & P. K., S. K. (2019). iPhone as a proxy indicator of adaptive narcissism: An empirical investigation. *Psychology & Marketing, 36*(10), 895–904. https://doi.org/10.1002/mar.21243

Santarossa, S., Coyne, P., Lisinski, C., & Woodruff, S. J. (2019). #fitspo on Instagram: A mixed-methods approach using Netlytic and photo analysis, uncovering the online discussion and author/image characteristics. *Journal of Health Psychology, 24*(3), 376–385.

Schrock, A. R. (2015). Communicative affordances of mobile media: Portability, availability, locatability, and multimediality. *International Journal of Communication, 9*, 1229–1246.

Shah, C. S., & Sebastian, M. P. (2020). Sentiment analysis and topic modelling of Indian government's Twitter Handle#IndiaFightsCorona. In *International Working Conference on Transfer and Diffusion of IT* (pp. 339–351). Springer: Cham.

Shaw, H., Ellis, D. A., Kendrick, L., Ziegler, F., & Wiseman, R. (2016). Predicting smartphone operating system from personality and individual differences. *Cyberpsychology, Behavior and Social Networking, 19*(12), 727–732. https://doi.org/10.1089/cyber.2016.0324

Smith, A., & Brenner, J. (2012). Twitter use 2012. *Pew Internet & American Life Project, 4*, 1–12.

Statista (2022, Dec 1). Smartphone penetration rate in India from 2010 to 2020, with estimates until 2040. Retrieved from www.statista.com/statistics/1229799/india-smartphone-penetration-rate/

Statista (2023, Jan 6). Share of Apple in the mobile phone market across India from January 2021 to December 2022. Retrieved from www.statista.com/statistics/938469/india-apple-share-in-the-mobile-phone-market/

Statista (2023b, Mar 8). Market share of mobile operating systems in India from 2012 to 2022. Retrieved from www.statista.com/statistics/262157/market-share-held-by-mobile-operating-systems-in-india/.

Twitter Marketing. (2013, 11 Feb). New compete study: Primary mobile users on Twitter. Retrieved from https://blog.twitter.com/en_us/a/2013/new-compete-study-primary-mobile-users-on-twitter

Wehner, M. R., Chren, M. M., Shive, M. L., Resneck, J. S., Pagoto, S., Seidenberg, A. B., & Linos, E. (2014). Twitter: An opportunity for public health campaigns. *The Lancet, 384*(9938), 131–132.

9 Health Information Seeking via WeChat, Social Determinants, and COVID-19 Vaccination Intentions

An Exploratory Study[1]

Li Chen, Yafei Zhang, and Ge Zhu

Introduction

In the past decade, health care agencies, health educators, and ordinary users across the globe have relied heavily on mobile apps to disseminate and acquire health information (Deng & Liu, 2017). In addition to turning to health apps for information about specific health topics (Fox & Duggan, 2012; Park, 2016), individuals have been utilizing a wide variety of mobile social media apps to obtain an even wider variety of health information and seek multi-dimensions of social support (Deng & Liu, 2017; Zhao et al., 2020). While health apps are specifically designed to monitor one's health status and to facilitate communication between healthcare providers and patients (Park, 2016), individuals use general social media apps like WeChat to achieve all kinds of interpersonal communication and social networking goals (e.g., Guo, 2017; Hou et al., 2018). Scholars have studied health information seeking via health apps intensively (e.g., Fox & Duggan, 2012; Park, 2016), but not much research has been done to examine health information seeking on social media apps. Health information seeking on mobile social media is likely to demonstrate unique features and generate new implications.

In China, WeChat is the most used social media app for health information (Zhang et al., 2017). During the COVID-19 pandemic, both the public and the private sectors intensively use mobile apps to disseminate health knowledge, update infection risk scores, and announce policy changes (Lu et al., 2021). While these approaches could mitigate the spread of the pandemic and reduce anxiety (Lu et al., 2021; Zhao et al., 2020), few studies have explored the demographic profile of users who actively seek such information or the effects of WeChat health information seeking. The present project aimed to address this research gap.

We chose to study WeChat because the app is drastically different from social media and microblogging sites such as Weibo. While Weibo creates a networked public space where information is open to all users by default (Han, 2018), WeChat creates a networked semiprivate space where users' existing personal networks determine the content they have access to (Guo, 2017). Our study provides new insights into the effects of the shift from *what you say* to *who you are* (Hou et al., 2018) on communication outcomes.

DOI: 10.4324/9781003328896-12

Instead of surveying various health attitudes and beliefs, we focused on those concerned with COVID-19 vaccination. Adopting the Structural Influence Model of Communication (SIM) framework, we examined WeChat users' health information seeking activities during the COVID-19 pandemic. Using an online survey, we investigated two major issues: (1) predictors of COVID-related health information seeking on WeChat and (2) the direct and indirect effects of social determinants, health information seeking, and vaccine beliefs on perceptions of the COVID-19 vaccines.

Literature Review

Structural Influence Model of Communication (SIM)

We choose the SIM as our main theoretical framework. In addition to the cognitive variables that directly predict health outcomes, social determinants, including sociodemographic, socio-economic, and place-related variables, play a crucial role in shaping health attitudes and behaviors (World Health Organization, n.d.). Public health scholars have adopted the social determinants of health approach to scrutinize the underlying sociocultural issues associated with vaccination (e.g., Gatwood et al., 2020; Thompson et al., 2019) and to reduce health disparities (Thompson et al., 2019).

SIM further elaborates on the effects of social determinants on communication and health outcomes. Several key components in SIM include individual social determinants, mediating or moderating conditions, health communication outcomes, and health outcomes (Bigsby & Hovick, 2018).

SIM has been proven to be an effective framework to examine how individuals' social determinants influence health communication and ultimately affect health outcomes (Bigsby & Hovick, 2018; Hovick et al., 2014). Modifying the model by Bigsby & Hovick (2018), we propose the following conceptual model (Figure 9.1).

SIM proposes that individuals' social determinants have a direct impact on their health communication outcomes (Bigsby & Hovick, 2018; Hovick et al., 2014).

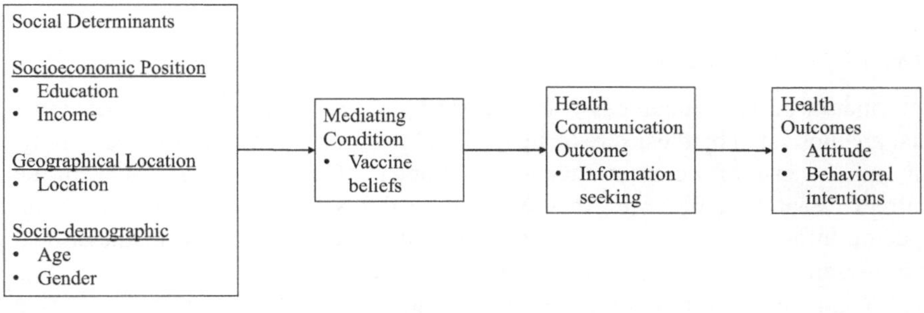

Figure 9.1 The Conceptual Model of Contextualized Structural Influence Model of Communication.

Existing literature suggests that varying determinants, such as age (Tan, 2015), gender (Qin et al., 2021), income, and education, are associated with individuals' health communication and behaviors (Bigsby & Hovick, 2018; Hovick et al., 2014; Kim & Jung, 2017; Tan, 2015). In this study, we focus on SES status (i.e., education and income), location, and socio-demographics (i.e., age and gender). Due to information asymmetry and the knowledge gap, individuals with low SES status are less likely to seek health information (Hovick et al., 2014; Sullivan & Rutten, 2009). Therefore, we argue that inequalities in SES could influence individuals' COVID-related information seeking. In line with previous research (e.g., Qin et al., 2021; Wang et al., 2020), this study also proposes a link between individuals' SES and their attitudes toward the COVID-19 vaccine as well as their vaccination intentions.

This study also considers geographical location as an important antecedent to communication and health outcome variables. In the United States, rural residents appear to be more vulnerable to the pandemic (Dobis & McGranahan, 2021) and vaccination coverage in rural areas is lower than in urban areas (Murthy et al., 2021). This trend is likely to extend to China, and it is probable that urban and rural Chinese residents react differently with regard to health information seeking and vaccination intentions during the pandemic.

Social determinants directly or indirectly influence vaccination intentions. A large-scale retrospective study examined global trends in vaccine uptake between 2015 and 2019. While results vary greatly by country, several consistent trends are shown across the globe: confidence in vaccines, trust in health care workers, higher levels of science education, being female, of a younger age, and high information seeking activities are strongly associated with vaccine uptake in most countries (Figueiredo et al., 2020). A recent survey of European adults suggests that COVID-19 vaccination intentions are associated with ethnicity, gender, education, annual income, knowledge of COVD-19, and adherence to government guidelines (Paul et al., 2021).

Synthesizing the above literature, we ask the following research question:

RQ1: How do social determinants influence (a) health information seeking; (b) attitude towards the COVID-19 vaccines; and (c) vaccination intentions?

Health Information Seeking

Information seeking encompasses a wide variety of active and intentional behaviors that are motivated by the desire to answer one's own questions, reduce uncertainty, and make sense of a current situation or problem (Case & Given, 2012). In the current research context, information seeking refers to "active efforts to obtain specific information outside of the normal patterns of exposure to mediated and interpersonal sources" (Niederdeppe et al., 2007, p. 155).

Typical health information seeking activities include using a search engine to find information about a particular health problem, reading blog posts about a health issue, and browsing health websites (Niederdeppe et al., 2007). Health information seeking is accounted for by gender: females are more likely to seek information about health or medical topics online than are males (Lustria et al, 2011; Waters et al., 2016). Another variable, age, is negatively associated with online health information seeking (Shim et al., 2006; Waters et al., 2016).

Moreover, the disparities in health information seeking is caused by socio-economic factors such as education, income, and place of residence. For example, people with lower educational levels tend not to use the Internet as their first source of health information, which could be explained by the positive correlation between education and health literacy (Shieh et al., 2009). Information on the Internet is rather complex and requires high reading levels to comprehend (Thompson et al., 2019), and self-efficacy barriers dissuade individuals with lower educational levels from actively seeking health information online (Shieh et al., 2009). In addition, income explains the variations of online health information seeking: low-income populations are less likely to use the Internet as their first source of health information (Reisdorf et al., 2021; Tang et al., 2019). Scholars also suggest that urban residents are more likely to obtain access to the Internet so that they more actively seek health information online than rural residents (Lustria et al., 2011; Reisdorf et al., 2021). Nevertheless, due to the relatively low cost of mobile phone adoption (Yan & Schroeder, 2020), lower income families and rural users (Yan & Schroeder, 2020) (Tsetsi & Rains, 2017) are no longer at a disadvantage regarding obtaining access to the Internet via a mobile phone. As a result, the divide between rural and urban communities in Internet access is shrinking, which may leading to increased online health information seeking by rural populations (Reisdorf et al., 2021).

Health information seeking has particular implications for vaccination. High health information seeking is associated with stronger vaccination intentions (Thompson et al., 2019) and improved vaccine uptake (Figueiredo et al., 2020). In the current study, we seek to reveal the direct and indirect effects of health information seeking via WeChat on users' attitudes toward the COVID-19 vaccines and vaccination intentions.

RQ2: How does health information seeking influence (a) attitude towards the COVID-19 vaccines and (b) vaccination intentions?

H1: Health information seeking mediates the effects of social determinants on (a) attitude towards the COVID-19 vaccines and (b) vaccination intentions.

Vaccine Beliefs

Existing and long-held vaccine beliefs shape individuals' perceptions and acceptance of emerging vaccines (e.g., Lazarus et al., 2020; Paul et al., 2021). Therefore, it is worthwhile to include participants' existing perceptions of vaccines and vaccination in our analysis framework.

The formation of vaccine beliefs is context-specific (e.g., Lazarus et al., 2020). The associations between social determinants and beliefs about the COVID-19 vaccines vary greatly across countries (Figueiredo et al., 2020). For example, education is positively associated with vaccine acceptance in some countries, but is negatively associated with vaccine acceptance in other countries (Lazarus et al., 2020). In Canada, Poland, France, Germany, and Sweden, older adults were more likely to accept the COVID-19 vaccine, while in China, younger people had more positive beliefs about the vaccine than older adults (Lazarus et al., 2020).

Built upon existing literature, we propose the following questions and hypothesis:

RQ3: How do vaccine beliefs influence (a) health information seeking; (b) attitude towards the COVID-19 vaccines; and (c) vaccination intentions?

RQ4: What is the role of vaccine beliefs as a mediator of the effects of social determinants on health information seeking?

H2: Health information seeking mediates the effects of vaccine beliefs on (a) attitude towards the COVID-19 vaccines and (b) vaccination intentions.

Methods

Participants

Active WeChat users who had not received the COVID-19 vaccines were invited to participate in this study. The online questionnaire was hosted by Qualtrics. To recruit a representative sample population, we utilized the paid services offered by Wenjuanxing (web address: wjx.com), a third-party crowdsourcing platform. Wenjuanxing has been proven to be a reliable platform to recruit research subjects in China (e.g., Men et al., 2021; Wang & Wu, 2021). Wenjuanxing distributed a link to the Qualtrics online questionnaire among its user pool. Each participant who completed the survey received 5.00RMB (about $0.77) in monetary compensation.

From July 8 to July 27, 2021, 521 individuals responded to the questionnaire. We removed the data points of participants who either failed to answer the attention check question correctly or left more than one third of the survey questions blank. After removing ineligible responses, a total of 405 responses were included in the dataset.

Primary Measures

Social Determinants. This study examined the following social determinants: age, gender, geographical location, education, and income. Table 9.1 displays participant demographics.

Vaccine Beliefs. In this study, vaccine beliefs refer to one's beliefs about vaccines and vaccination in general, not limited to the COVID-19 vaccines. This variable was measured using twelve 7-point Likert scale items (e.g., "I feel protected after getting vaccinated") from the Vaccine Attitudes Examination Scale (Martin & Petrie, 2017). With the completion of reverse coding, higher scores indicated more positive vaccine beliefs, $\alpha = .87$, $M = 4.62$, $SD = 0.96$.

Table 9.1 Participant Demographics

Age		
	$M = 30.04$	
	$SD = 7.58$	
Gender		
	Male	$n = 198$ (48.9%)
	Female	$n = 207$ (51.1%)
Geographical location		
	Urban	$n = 347$ (85.7%)
	Rural	$n = 16$ (4.0%)
	Urban-rural	$n = 41$ (10.1%)
	Other	$n = 1$ (0.2%)
Education		
	Below high school	$n = 7$ (1.7%)
	Secondary vocational school	$n = 11$ (2.7%)
	High school	$n = 24$ (5.9%)
	Undergraduate student	$n = 61$ (15.1%)
	Bachelor's degree	$n = 270$ (66.7%)
	Master's degree and above	$n = 32$ (7.9%)
Annual income (RMB)		
	Less than 10,000	$n = 55$ (13.6%)
	10,000 – 49,999	$n = 49$ (12.1%)
	50,000 – 99,999	$n = 97$ (24.0%)
	100,000 – 149,999	$n = 100$ (24.7%)
	150,000 – 199,999	$n = 58$ (14.3%)
	200,000 – 249,999	$n = 28$ (6.9%)
	250,000 – 300,000	$n = 9$ (2.2%)
	More than 300,000	$n = 9$ (2.2%)

Source: Table created by authors.

Health Information Seeking. The measures of health information seeking were adapted from Hovick & Bigsby (2016). Participants answered a set of questions: "Within the past 30 days, how much have you actively looked for information about the COVID-19 vaccine from [information source]?" In this study, information source included individual WeChat friends (M = 3.08, SD = 1.46), WeChat Moments (M = 2.90, SD = 1.47), WeChat group chats (M = 2.89, SD = 1.50), official accounts by governmental sectors (M = 3.55, SD = 1.59), official accounts by public health experts (M = 3.03, SD = 1.73), official accounts by celebrities (M = 2.07, SD = 1.34), and official accounts by hospitals (M = 2.96, SD = 1.63). The values of health information seeking ranged from 1 [not at all] to 7 [very frequently].

Attitude Towards the COVID-19 Vaccines. Adapted from the works of Ajzen (1991) and Boudewyns and Paquin (2011), six 7-point semantic differential scale questions measured attitude towards COVID-19 vaccination (e.g., "I think that getting a COVID-19 vaccine in the near future is [worthless – worthwhile]"), $\alpha = .94$, $M = 5.56$, $SD = 1.33$.

Vaccination Intentions. Adapting the 7-point semantic differential scale questions from Ajzen (1991) and Boudewyns and Paquin (2011), we used four questions to measure COVID-19 vaccination intentions (e.g., "How willing are you to get a COVID-19 vaccine in the near future?"), $\alpha = .96$, $M = 5.17$, $SD = 1.59$.

Analyses

Means, standard deviations, and correlations of independent variables and dependent variables are shown in Table 9.2. To test direct effects (RQs1-3), we conducted ordinary least squares (OLS) regression. To test indirect effects (H1, H2, and RQ4), the bootstrapping technique with 5,000 samples was used, which generated a 95 percent bias-corrected confidence interval for indirect effects (Bigsby & Hovick, 2018). We used multiple models with different dependent variables in relevant R packages (e.g., Lavaan). VIF ranged from 1.03 to 1.52, which excluded the multicollinearity issue in this study.

Results

Direct Effects

RQ1 asks about the associations between social determinants and (a) health information seeking activities on WeChat; (b) attitude towards the COVID-19 vaccines; and (c) vaccination intentions. Table 9.3 shows OLS regression results.

Gender and income are significant predictors of health information seeking (RQ1a). Compared to males, females are significantly more likely to actively seek COVID-related information on WeChat [$M_{male} = 2.69$, $SD_{male} = 1.17$, $M_{female} = 3.15$, $SD_{female} = 1.27$, $t(401) = 3.81$, $p <.000$]. Income is also positively associated with health information seeking: those who have a higher income seek COVID-related information on WeChat more actively [$\beta = .18$, $R^2 = .06$, $F(1, 401) = 23.70$, $p < .000$].

There is a weak and negative association between education and attitude towards the COVID-19 vaccines (RQ1b). Those who have higher educational backgrounds are less likely to form favorable attitudes toward COVID-19 vaccination [$\beta = -.19$, $R^2 = .02$, $F(1, 392) = 7.00$, $p = .008$].

No significant associations are found between social determinants and vaccination intentions (RQ1c).

Health information seeking has a positive direct effect on attitude towards the COVID-19 vaccines (RQ2a) and vaccination intentions (RQ2b). Active health information seeking results in more favorable attitudes toward the COVID-19 vaccine [$\beta = .27$, $R^2 = .06$, $F(1, 390) = 26.82$, $p < .000$] and stronger vaccination intentions [$\beta = .44$, $R^2 = .12$, $F(1, 400) = 53.92$, $p < .000$].

Table 9.2 Correlations, Means, and Standard Deviations of Study Variables

	Mean	*SD*	*Age*	*Location*	*Gender*	*Education*	*Income*	*Vaccine beliefs*	*Health information seeking*	*Attitude towards the COVID-19 vaccines*	*Vaccination intentions*
Age	30.04	7.58	-								
Location	-	-	−.05	-							
Gender	-	-	.10	−.02	-						
Education	4.66	0.93	.06	−.21***	−.16***	-					
Income	-	-	.35***	−.21***	−.01	.41***	-				
Vaccine beliefs	4.62	0.96	−.01	−.03	.09	−.03	.06	-			
Health information seeking	2.92	1.24	−.02	−.05	−.19***	.22***	.24***	.22***	-		
Attitude toward the COVID-19 vaccines	5.56	1.33	−.08	.02	.08	−.13***	−.04	.68***	.25***	-	
Vaccination intentions	5.17	1.59	−.07	−.02	.05	.04	.05	.54***	.34***	.72***	-

Source: Table created by authors.

Note. N = 405. Pearson correlations, two-tailed significance, listwise deletion. Means and standard deviations are only available for continuous variables or variables treated as continuous. $^{*}p < .05$, $^{**}p < .01$, $^{***}p < .001$.

Table 9.3 Predictors of Health Information Seeking, Attitude, and Vaccination Intentions

Independent variables	*Communication behavior*	*Dependent variables*	
	Health information seeking	*Attitude towards the COVID-19 vaccines*	*Vaccination intentions*
	b (SE)	*b (SE)*	*b (SE)*
Age	−.01 (.01)	−.01 (.01)	−.01 (.01)
Location	.04 (.18)	.05 (.15)	.01 (.20)
Gender	−.46 (.12)***	.12 (.10)	.17 (.14)
Education	.11 (.07)	−.13 (.06)*	.004 (.08)
Income	.17 (.04)***	−.04 (.04)	−.002 (.05)
Vaccine beliefs	.30 (.06)***	.89 (.05)***	.79 (.07)***
Health Information seeking		.17 (.04)***	.31 (.06)***
F	11.58***	53.41***	28.2***
R^2	.16	.50	.35

Source: Table created by authors.

Note. N = 405. Unstandardized ordinary least squares (OLS) regression coefficients are reported. *$p < .05$, **$p < .01$, ***$p < .001$.

RQ3 asks if vaccine beliefs predict the dependent variables. Vaccine beliefs are positively associated with health information seeking (RQ3a). The more supportive of vaccines and vaccination individuals are, the more actively they seek COVID-related information on WeChat [$\beta = .30$, $R^2 = .05$, $F(1, 392) = 19.26$, $p < .000$]. Vaccine beliefs are also a positive predictor of attitude towards the COVID-19 vaccines (RQ3b) and vaccination intentions (RQ3c). Those who have positive perceptions of vaccines tend to form more favorable attitudes toward COVID-19 vaccination [$\beta = .94$, $R^2 = .47$, $F(1, 382) = 335.26$, $p < .000$] and develop stronger intentions to get vaccinated [$\beta = .90$, $R^2 = .29$, $F(1, 392) = 162.84$, $p < .000$].

The above results show that three out of the five major demographic variables in our proposed conceptual model, including gender, income, and education, more or less predict health communication behaviors and outcomes, while age and geographical location are not significant predictors. In addition, compared to demographics, two psychographic variables, health information seeking and vaccine beliefs, exert a stronger influence on attitude towards the COVID-19 vaccines and vaccination intentions.

Indirect Effects

H1 predicts that health information seeking mediates the effects of social determinants on attitude and vaccination intentions. This hypothesis is not supported. No significant indirect effects are identified.

H2 predicts that health information seeking mediates the effects of vaccine beliefs on outcome variables. This hypothesis is supported. Individuals' vaccine

Table 9.4 Indirect Effects of Social Determinants on Outcomes via Health Information Seeking

Independent variables	*Mediator*	*Dependent variables*			
		Attitude towards the COVID-19 vaccines		*Vaccination intentions*	
		Est. (SE)	*95% BC*	*Est. (SE)*	*95% BC*
Age	Health information seeking	−.001 (.001)	[−.004, .001]	−.004 (.003)	[−.01, .002]
Location	Health information seeking	.008 (.02)	[−.04, .06]	.004 (.06)	[−.11, .13]
Gender	Health information seeking	.02 (.02)	[−.01, .06]	.05 (.05)	[−.03, .15]
Education	Health information seeking	−.02 (.01)	[−.05, −.002]	.001 (.03)	[−.05, .05]
Income	Health information seeking	−.01 (.01)	[−.02, .003]	−.001 (.02)	[−.04, .03]
Vaccine beliefs	Health information seeking	.15 (.04)	[.07, .22] ***	.24 (.05)	[.16, .34] ***

Source: Table created by authors.

Note. *$p < .05$, **$p < .01$, ***$p < .001$. BC 95% CI: Bias-corrected 95% bootstrapped confidence interval (CI) based on 5,000 resamples.

beliefs enhance their health information seeking, which, subsequently, results in more positive attitudes and stronger vaccination intentions (Table 9.4).

RQ4 asks if and how vaccine beliefs mediate the effects of social determinants on health information seeking. Vaccine beliefs mediate the relationship between two social determinants (gender and income) and health information seeking. Income positively predicts vaccine beliefs, which further leads to more active health information seeking (Table 9.5). The aforementioned indirect effects suggest that vaccine beliefs enlarge the effects of demographics on health communication outcomes.

Discussion

Health Information Seeking on WeChat

Empirical research in multiple social and cultural contexts has identified several common predictors of online health information seeking, including age, gender, education, income, Internet use abilities, and health literacy (Lustria et al., 2011; Reisdorf et al., 2021; Shieh et al., 2009; Shim et al., 2006; Thompson et al., 2019).

In our study, while education is not a significant predictor of health information seeking, income remains a positive predictor. Vaccine beliefs mediate the effects of

Table 9.5 Indirect Effects of Social Determinants on Health Information Seeking via Vaccine Beliefs

Independent variables	*Mediator*	*Dependent variable: Health information seeking*	
		Est. (SE)	*95% BC*
Age	Vaccine beliefs	−.004 (.003)	[−.01, .001]
Location	Vaccine beliefs	.01 (.05)	[−.09, .11]
Gender	Vaccine beliefs	−.14 (.05)	[−.24, −.06]**
Education	Vaccine beliefs	.03 (.02)	[−.004, .08]
Income	Vaccine beliefs	.05 (.02)	[.02, .09]**

Source: Table created by authors.

Note. $^*p < .05$, $^{**}p < .01$, $^{***}p < .001$. BC 95% CI: Bias-corrected 95% bootstrapped confidence interval (CI) based on 5,000 resamples.

gender and income on health information seeking. Both results resonate with the research findings by Tang et al. (2019). Among urban Chinese residents, income is positively associated with health information seeking and health literacy, resulting in better health outcomes (Tang et al., 2019). Our study shows that the trend extends to health information seeking on WeChat. Even though health literacy and vaccine beliefs are different variables, they are somewhat inherently connected: health literacy encompasses a collection of cognitive and analytical knowledge and skills (Baur, 2010), and vaccine beliefs can be one outcome of these knowledge and skills.

Early studies on the digital divide suggest that urban residents have more access to the Internet, so they more actively engage in online health information seeking (Lustria et al., 2011), resulting in positive health outcomes such as higher vaccination acceptance rates (e.g., Murthy et al., 2021). In our study, however, urban and non-urban residents have identical performances regarding health information seeking, attitude, and vaccination intentions.

In China, Internet access was 23 times lower in rural areas than in urban areas at the beginning of the twenty-first century (Fong, 2009), but the adoption of mobile devices has largely reduced the urban-rural gap in recent years (Yan & Schroeder, 2020). Due to the relatively low cost of mobile phone adoption (Fong, 2009; Yan & Schroeder, 2020), 84 percent of rural users access the Internet via mobile devices, "as opposed to 79 percent among urban users" (Yan & Schroeder, 2020, p. 319). Therefore, urban and rural users are almost at the same "starting point" regarding mobile media adoption. It is not surprising that geographical location is no longer a significant predictor of communication activities on WeChat. An identical trend has been identified among lower income families in the United States (Tsetsi & Rains, 2017). Our research findings resonate with the idea that smartphones and mobile apps may be a key player in narrowing the digital gap (e.g., Tsetsi & Rains, 2017).

Even though age was noted as a common predictor of health communication behaviors and outcomes (e.g., Tan, 2015), in our study the variable did not predict any dependent variables. One possible explanation is, while narrowing the digital gap, the proliferation of mobile devices could have blurred the distinctions between age groups. Older adults actively use WeChat to achieve a wide variety of purposes (Guo, 2017) so that the differences in media use habits between younger and older adults are no longer distinct.

Predictors of Vaccination Intentions

Negative attitudes toward vaccines and unwillingness to get vaccinated have become major barriers to curbing the COVID-19 pandemic (Paul et al., 2021). In China, vaccine hesitancy is rising (Yang et al., 2020). Individuals began to question the reliability of vaccines after a series of vaccine safety scandals in the past two decades (Yang et al., 2020).

Nevertheless, just as vaccine opponents tend to question all kinds of vaccines (Lewandowsky et al., 2012), those who hold positive vaccine beliefs are inclined to trust and accept new vaccines. The trend shows that one's existing vaccine beliefs are persistent enough to influence one's attitude towards and intentions to get an emerging vaccine. Therefore, cultivating positive vaccine beliefs and rebuilding public trust in vaccinations benefit society over the long term by enhancing public trust in emerging medical interventions, like COVID-19 vaccination, during outbreaks of emerging infectious diseases.

Health information seeking mediates the effects of vaccine beliefs on attitudes towards the COVID-19 vaccines and vaccination intentions. The mediation effects can be explained by confirmation bias. Individuals tend to select belief-consistent messages in the media and perceive belief-consistent information as credible and convincing (Meppelink et al., 2019). Unlike conventional mass media such as TV and newspapers that deliver mass messages to large audiences, mobile apps like WeChat allow users to follow the individuals and accounts they want to follow and transition their offline social networks online (Guo, 2017). Therefore, those who hold positive vaccine beliefs are likely to follow individuals and accounts that disseminate pro-vaccination messages, such as public health agencies (Zhang et al., 2019). Since these messages are consistent with their existing beliefs, they tend to trust them, further enhancing their trust in COVID-19 vaccination. Mobile communication tends to reinforce one's existing viewpoints and perceptions rather than changing them.

The mediation effects of health information seeking also imply that WeChat use is likely to lower vaccine opponents' trust in the COVID-19 vaccine and weaken their vaccination intentions. Unlike Twitter and Weibo, which foster conversations open to the public (Jia, 2017), WeChat facilitates a semiprivate networked space (Guo, 2017) in which users' existing relationships determine the media content to which they have access. Therefore, vaccine opponents are likely to join vaccine resistance groups and follow public accounts that disseminate anti-vaccination messages. The semiprivate spaces on WeChat could have enlarged the gap between vaccine supporters and opponents. This new form of the digital divide is not caused by socio-economic backgrounds (van Deursen & van Dijk, 2010; Zhao et al., 2020), but is facilitated by a mobile app. With the proliferation of the Internet and digital media, psychographics such as health literacy (Quinn et al., 2017), navigational needs (Lee et al., 2015), and media-use skills (van Deursen & van Dijk, 2010) play an increasingly prominent role in enlarging the digital gap between social groups. Our study suggests that one's existing social network can be a new factor in widening the gap. In this case, mobile apps like WeChat enlarge the discrepancies between individuals who hold distinct vaccine beliefs, which is worthwhile to note in health education and advocacy.

Conclusion

Within the SIM framework, we utilized an online survey to investigate WeChat users' health information seeking activities during the COVID-19 pandemic. Three major findings were identified.

First, gender, income, and vaccine beliefs are positively associated with health information seeking. Health information seeking, subsequently, has a positive effect on attitude towards the COVID-19 vaccines and vaccination intentions. Health information seeking also mediates the effects of vaccine beliefs on outcome variables. Second, vaccine beliefs mediate the effects of gender and income on health information seeking. Finally, while vaccine beliefs are a positive predictor of attitude and intentions, educational background is a negative predictor of attitude.

Introducing vaccine beliefs and health information seeking to SIM connects SIM with other widely used health information seeking models such as the Planned Risk Information Seeking Model (PRISM) and Situational Theory of Problem Solving (STOPS). These models categorize predictors of health information seeking into four types, namely psychological factors, instrumental factors, contextual factors, and demographic factors (Wang et al., 2021). Our study shows that vaccine beliefs can serve as a new psychosocial factor and health information seeking is a contextual factor to predict vaccine-related information seeking.

In addition, our study suggests that compared to demographics, psychographics play a more critical role in shaping communication activities on mobile social media apps. Therefore, when applying a typical health communication theory to mobile communication, scholars are encouraged to place more emphasis on psychographics, especially attitudes, beliefs, and behavioral intentions.

Regarding practical implications, empirical research in various contexts suggests that health information seeking empowers individuals in health decision-making (Manafo & Wong, 2012), alters their lifestyle behaviors (Kelly et al., 2010; Shim et al., 2006), and encourages vaccine uptake (Figueiredo et al., 2020; Jung et al., 2013; Kim & Jung, 2017; Thompson et al., 2019). Our study confirms the positive effects of health information seeking on attitude towards the COVID-19 vaccine and vaccination intentions. Given WeChat's popularity among Chinese users, health educators are encouraged to more intensively use mobile apps like WeChat to reach out to large and diverse audience segments, disseminate reliable health information, and build a relationship of trust between ordinary users and the scientific community.

This study has several limitations. In our data analysis, we did not differentiate between different information sources, such as WeChat Moments and the official accounts of public health experts. The quality of health information from different sources and individuals' trust in different sources vary greatly (e.g., Waters et al., 2016). Both factors are likely to shape health communication outcomes. Future research could treat each health information source as a separate variable and compare the effects of different information sources. In addition, this study only examined the frequency of health information seeking but did not take into consideration motivations for health information seeking. Future research could study various health information seeking motivations (Case & Given, 2012) to scrutinize the nuances of health communication on mobile social media.

Note

1 The article is revised based on the article published in *China Media Research* Volume 18, Issue 1, 2022, pp. 1–22. Reproduced with permission from *China Media Research.*

References

Ajzen, I. (1991). The theory of planned behavior. *Organizational Behavior and Human Decision Processes, 50*(2), 179–211.

Baur, C. (2010). New directions in research on public health and health literacy. *Journal of Health Communication, 15*, 42–50.

Bigsby, E., & Hovick, S. R. (2018). Understanding associations between information seeking and scanning and health risk behaviors: an early test of the structural influence model. *Health Communication, 33*(3), 315–325.

Boudewyns, V., & Paquin, R. S. (2011). Intentions and beliefs about getting tested for STDs: implications for communication interventions. *Health Communication, 26*(8), 701–711.

Case, D. O., & Given, L. M. (2012). *Looking for information: a survey of research on information seeking, needs, and behavior (4th ed.).* Emerald Publishing Group.

Deng, Z., & Liu, S. (2017). Understanding consumer health information-seeking behavior from the perspective of the risk perception attitude framework and social support in mobile social media websites. *International Journal of Medical Informatics, 105*, 98–109.

Dobis, E. A., & McGranahan, D. (2021, February 1). Rural residents appear to be more vulnerable to serious infection or death from coronavirus COVID-19. www.ers.usda.gov/amber-waves/2021/february/rural-residents-appear-to-be-more-vulnerable-to-serious-infection-or-death-from-coronavirus-covid-19/

Figueiredo de, A., Simas, C., Karafillakis, E., Paterson, P., & Larson, H. J. (2020). Mapping global trends in vaccine confidence and investigating barriers to vaccine uptake: A large-scale retrospective temporal modelling study. *The Lancet, 396*(10255), 898–908.

Fong, M. W. L. (2009). Digital divide between urban and rural regions in China. *The Electronics Journal on Information Systems in Developing Countries, 36*(6), 1–12.

Fox, S., & Duggan, M. (2012). Mobile health 2012. *Pew Research Center's Internet & American Life Project.* http://pewinternet.org/Reports/2012/Mobile-Health.aspx

Gatwood, J., Shuvo, S., Hohmeier, K. C., Hagemann, T., Chiu, C.-Y., Tong, R., & Ramachandran, S. (2020). Pneumococcal vaccination in older adults: An initial analysis of social determinants of health and vaccine uptake. *Vaccine*, *38*(35), 5607–5617.

Guo, L. (2017). WeChat as a semipublic alternative sphere: exploring the use of WeChat among Chinese older adults. *International Journal of Communication*, *11*, 408–428.

Han, E. L. (2018). Weibo and the making of Chinese networked publics: witness, debates and expertise. *Communication and the Public*, *3*(2), 97–112.

Hou, J., Ndasauka, Y., Pan, X., Chen, S., Xu, F., & Zhang, X. (2018). Weibo or WeChat? Assessing preference for social networking sites and role of personality traits and psychological factors. *Frontiers in Psychology*, *9*. doi: 10.3389/fpsyg.2018.00545

Hovick, S. R., & Bigsby, E. (2016). Heart disease and colon cancer prevention beliefs and their association with information seeking and scanning. *Journal of Health Communication*, *21*(1), 76–84.

Hovick, S. R., Liang, M., & Kahlor, L. (2014). Predicting cancer risk knowledge and information seeking: the role of social and cognitive factors. *Health Communication*, *29*(7), 656–668.

Jia, H., Wang, D., Miao, W., & Zhu, H. (2017). Encountered but not engaged: examining the use of social media for science communication by Chinese scientists. *Science Communication*. doi: 10.1177/1075547017735114

Jung, M., Lin, L., & Viswanath, K. (2013). Associations between health communication behaviors, neighborhood social capital, vaccine knowledge, and parents' H1N1 vaccination of their children. *Vaccine*, *31*(42), 4860–4866.

Kelly, B., Hornik, R., Romantan, A., Schwartz, J. S., Armstrong, K., DeMichele, A., Fishbein, M., Gray, S., Hull, S., Kim, A., Nagler, R., Niederdeppe, J., Ramirez, A. S., Smith-McLallen, A., & Wong, N. (2010). Cancer information scanning and seeking in the general population. *Journal of Health Communication*, *15*(7), 734–753.

Kim, J., & Jung, M. (2017). Associations between media use and health information-seeking behavior on vaccinations in South Korea. *BMC Public Health*, 17. doi: 10.1186/s12889-017-4721-x

Lazarus, J. V., Wyka, K., Rauh, L., Rabin, K., Ratzan, S., Gostin, L. O., Larson, H. J., & El-Mohandes, A. (2020). Hesitant or not? The association of age, gender, and education with potential acceptance of a COVID-19 vaccine: a country-level analysis. *Journal of Health Communication*, *25*(10), 799–807.

Lee, K., Hoti, K., Hughes, J. D., & Emmerton, L. M. (2015). Consumer use of "Dr Google": a survey on health information-seeking behaviors and navigational needs. *Journal of Medical Internet Research*, *17*(12), e288.

Lewandowsky, S., Ecker, U. K. H., Seifert, C. M., Schwarz, N., & Cook, J. (2012). Misinformation and its correction: continued influence and successful debiasing. *Psychological Science in the Public Interest*, *13*(3), 106–131.

Lu, Q., Jiang, W., Zhang, X., Li, H., Zhang, X., Zeng, H., Du, J., Yang, G., Zhang, L., Li, R., Fang, L., Li, H., & Liu, W. (2021). Mobile health technology combats COVID-19 in China [letter to the editor]. *Journal of Infection*, *82*, 167–168.

Lustria, M. L., Smith, S. A., & Hinnant, C. C. (2011). Exploring digital divides: an examination of ehealth technology use in health information seeking, communication and personal health information management in the USA. *Health Informatics Journals*, *17*(3), 224–243.

Manafo, E., & Wong, S. (2012). Exploring older adults' health information seeking behaviors. *Journal of Nutrition Education and Behavior*, *44*(1), 85–89.

Martin, L. R., & Petrie, K. J. (2017). Understanding of dimensions of anti-vaccination attitudes: the vaccination attitudes examination (VAX) scale. *Annals of Behavioral Medicine*. doi: 10.1007/s12160-017-9888-y.

Men, L. J., Qin, Y. S., & Mitson, R. (2021). Engaging startup employees via charismatic leadership communication: the importance of communicating "vision, passion, and care." *International Journal of Business Communication*, 1–21. https://doi.org/10.1177/23294884211020488

Meppelink, C. S., Smit, E. G., Fransen, M. L., & Diviani, N. (2019). "I was right about vaccination": confirmation bias and health literacy in online health information seeking. *Journal of Health Communication, 24*(2), 129–140.

Murthy, B. P., Sterrett, N., Weller, D., Zell, E., Reynolds, L., Toblin, R. L., Murthy, N., Kriss, J., Rose, C., Cadwell, B., Wang, A., Ritchey, M. D., Gibbs-Scharf, L., Qualters, J. R., Shaw, L., Brookmeyer, K. A., Clayton, H., Eke, P., Adams, L., Harris, L. Q. (2021, May 21). Disparities in COVID-19 vaccination coverage between urban and rural counties – United States, December 14, 2020 – April 10, 2021. *Morbidity and Mortality Weekly Report*. www.cdc.gov/mmwr/volumes/70/wr/mm7020e3.htm#suggestedcitation

Niederdeppe, J., Hornik, R. C., Kelly, B. J., Frosch, D. L., Romantan, A., Stevens, R. S., Barg, F. K., Weiner, J. L., & Schwartz, J. S. (2007). Examining the dimensions of cancer-related information seeking and scanning behavior. *Health Communication, 22*(2), 153–167.

Park, Y. (2016). Emerging new era of mobile health technologies. *Healthcare Informatics Research, 22*(4), 253–254.

Paul, E., Steptoe, A., & Fancourt, D. (2021). Attitudes towards vaccines and intention to vaccinate against COVID-19: implications for public health communications. *The Lancet Regional Health – Europe, 1*. https://doi.org/10.1016/j.lanepe.2020.100012

Qin, W., Wang, E., & Ni, Z. (2021). Chinese consumers' willingness to get a COVID-19 vaccine and willingness to pay for it. *Plos One, 16*(5), e0250112.

Quinn, S., Bond, R., & Nugent, C. (2017). Quantifying health literacy and eHealth literacy using existing instruments and browser-based software for tracking online health information seeking behavior. *Computers in Human Behavior, 69*, 256–267.

Reisdorf, B., Blank, G., Bauer, J. M., Cotten, S. R., Robertson, C., & Knittel, M. (2021). Information-seeking patterns and COVID-19 in the United States. *Journal of Quantitative Description: Digital Media, 1*, 1–38.

Shieh, C., Mays, R., McDaniel, A., & Yu, J. (2009). Health literacy and its association with the use of information sources and with barriers to information seeking in clinic-based pregnant women. *Health Care for Women International, 30*(11), 971–988.

Shim, M., Kelly, B., & Hornik, R. (2006). Cancer information scanning and seeking behavior is associated with knowledge, lifestyle choices, and screening. *Journal of Health Communication, 11*, 157–172.

Sullivan, H. W., & Rutten, L. J. F. (2009). Cancer prevention information seeking: a signal detection analysis of data from the cancer information service. *Journal of Health Communication, 14*(8), 785–796.

Tan, A. S. L. (2015). A study of the frequency and social determinants of exposure to cancer-related direct-to-consumer advertising among breast, prostate, and colorectal cancer patients. *Health Communication, 30*(11), 1102–1111.

Tang, C., Wu, X., Chen, X., Pan, B., & Yang, X. (2019). Examining income-related inequality in health literacy and health-information seeking among urban population in China. *BMC Public Health, 19*(1), 1–9.

Thompson, E. L., Rosen, B. L., & Maness, S. B. (2019). Social determinants of health and human papillomavirus vaccination among young adults, national health interview survey 2016. *Journal of Community Health*, *44*(1), 149–158.

Tsetsi, E., & Rains, S. A. (2017). Smartphone Internet access and use: extending the digital divide and usage gap. *Mobile Media & Communication*, *5*(3), 239–255.

Van Deursen, A., & Van Dijk, J. (2010). Internet skills and the digital divide. *New Media & Society*, *13*(6), 893–911.

Wang, J., Jing, R., Lai, X., Zhang, H., Lyu, Y., Knoll, M. D., & Fang, H. (2020). Acceptance of COVID-19 vaccination during the COVID-19 pandemic in China. *Vaccines*, *8*(3), 482.

Wang, W., & Wu, J. (2021). Short video platforms and local community building in China. *International Journal of Communication*, *15*, 3269–3291.

Wang, X., Shi, J., & Kong, H. (2021). Online health information seeking: a review and meta-analysis. *Health Communication*, *36*(10), 1163–1175.

Waters, E. A., Wheeler, C., & Hamilton, J. G. (2016). How are information seeking, scanning, and processing related to beliefs about the roles of genetics and behavior in cancer causation? *Journal of Health Communication*, *21*, 6–15.

World Health Organization. (n.d.). *Social determinants of health*. www.who.int/social_determinants/en

Yan, P., & Schroeder, R. (2020). Variations in the adoption and use of mobile social apps in everyday lives in urban and rural China. *Mobile Media and Communication*, *8*(3), 318–341.

Yang, R., Penders, B., & Horstman, K. (2020). Addressing vaccine hesitancy in China: a scoping review of Chinese scholarship. *Vaccines*, *8*, 2–17.

Zhang, X., Wen, D., Liang, J., & Lei, J. (2017). How the public uses social media WeChat to obtain health information in China: a survey study. *BMC Medical Informatics and Decision Making*, *17*, 71–79.

Zhang, Y., Xia, T., Huang, L., Yin, M., Sun, M., Huang, J., Ni, Y., & Ni, J. (2019). Factors influencing user engagement of health information disseminated by Chinese provincial Centers for Disease Control and Prevention on WeChat: observational study. *JMIR mHealth and uHealth*, *7*(6), e12245. doi: 10.2196/12245

Zhao, X., Fan, J., Basnyat, I., & Hu, B. (2020). Online health information seeking using "#COVID-19 Patient Seeking Help" on Weibo in Wuhan, China: descriptive study. *Journal of Medical Internet Research*, *22*(10), e22910. doi: 10.2196/22910

10 Mobile Communication and New Social Movement

Evidence from Global South

Shudipta Sharma

Introduction

With the increasing number of users, mobile media occupies a significant place in social, political, and economic reconfiguration across the world. The portability feature of the new media, which allows mobile communication to overcome time and space limitations, contributes to the growing mobile network society (Siapera, 2012). Studies also find mobile media, more specifically mobile phones and wireless Internet, have become major tools in contemporary social movements (Castells, 2007; Udupa et al., 2020). Besides initiating, organizing, and coordinating a movement, mobile communication plays crucial roles in leadership and decision-making processes (Sharma, 2020). On several occasions, online activism led to offline social movements (Harlow, 2012). Many offline movements also attracted the attention of mobile media users who extended those movements across the country and beyond (Sharma, 2014; Rahman & Hasan, 2022). From the Arab Spring to the Occupy Wall Street movements, these new trends have become visible in the protests that spread beyond borders. Mobile communication plays similar roles to the protests limited to a country or city (i.e., Gezi Park Resistance and the Umbrella Revolution).

Bangladesh, a South Asian country, is not an exception. New information and communication technologies (ICTs) are widely used in every recent movement in this developing country. However, most of the literature on mobile communication in the region deals with economic growth and development (e.g., Rahman et al., 2020; Sharma et al., 2022), although numerous studies on other parts of the world focus on cultural and social aspects as well (Castells et al., 2007; Katz & Arkus, 2002). This chapter aims to address this gap. More specifically, it contributes to the growing literature by looking at how mobile communication is used for inception, mobilization, and organization in social movements in the Global South. This study also aims to understand how mobile communication leads an online movement to an offline one in a region with low socioeconomic status and limited access to ICTs. To do so, this qualitative study explores the Shahbag movement in Bangladesh. The movement was initiated by some "unknown bloggers" in 2013, demanding capital punishment for war criminals who assisted the Pakistan Army during the Liberation War of Bangladesh in 1971. Although only 6.7 percent of

DOI: 10.4324/9781003328896-13

Bangladeshis had Internet access at that time, people from all walks of life joined the movement and organized the largest mass gathering the nation had seen in 20 years (Zaman, 2016). The movement is considered a milestone in the politics of Bangladesh since it not only rediscovered the prospects of mobile communication in politics but also redefined the political landscape of the country (Roy, 2019).

The chapter is divided into six sections. The first section discusses the significance and objectives of the study. Besides briefly describing the socioeconomic and the ICTs status of Bangladesh; the second section deals with contextualizing the Shahbag movement. The theoretical framework and existing literature on the relationship between social movements and mobile communication are reviewed in the third section. After discussing the methodology in the fourth section, the findings of the study are introduced in the fifth section with a focus on the role of mobile communication in initiating, mobilizing, and organizing the Shahbag movement. Discussion and concluding remarks are made in the final section, where limitations and suggestions for future research are also discussed.

Contextualizing the Shahbag Movement

Bangladesh has a population of over 165 million, making it one of the most densely populated countries in the world. A significant majority of the population (70%) lives in rural areas, while the rest reside in towns and cities. The country has 45.9 million young people aged between 15 and 29, accounting for 27.82 percent of the total population. Over the years, the literacy rate in the country has significantly increased, standing at 74.66 percent in 2022, up from 51.77 percent in 2011 (BBS, 2022). The use of mobile phones, the Internet, and social media is also growing rapidly in Bangladesh. The country is the fifth-largest mobile market in the Asia Pacific region (*The Business Standard*, 2020), with more mobile connections than the population, and 49.55 million social media users, representing 29.7 percent of the population (Kemp, 2022). About 52.2 percent of people use smartphones, while 38.9 percent have access to the Internet (*The Business Standard*, 2022). Facebook is the most popular social media platform, with 58.94 million people using it, accounting for 35.3 percent of the population (Internet World Stats, 2022). Social media is found to be used for both constructive purposes such as citizen journalism and election, allowing individuals to share news and information (Sharma & Rahaman, 2018), as well as for destructive purposes including the promotion of violent extremism and other harmful activities (Sharma, 2015; Sharma, 2023). While the socioeconomic conditions of the country have improved and Bangladesh has become a lower-middle-income country recently, around 39 million people still live below the national poverty line (World Bank, 2020). Moreover, when the Shahbag movement was organized in 2013, 64 percent of people used mobile phones, while only 6.7 percent had Internet access (*The Business Standard*, 2022), and most people were not familiar with the term "blogger" when the literacy rate was 57 percent (Hasan, 2014). It is also found that social media literacy is very low among the youths in the country (Sharma & Islam, 2022).

The Shahbag movement was initiated by bloggers and young cultural and political activists demanding capital punishment for war criminals of the Liberation War of Bangladesh in 1971. The Pakistan Army, aided by its local collaborators, including Islamist militias from Jamaat-e-Islami (JI), killed three million people, raped 200,000 women, and forced 10 million people to flee to refugee camps during the nine-month-long war (Bass, 2013; Sharma & Ahsan, 2015). Although the first government of the country took some initiatives to arrange the trial of war criminals after the independence in 1971, they could not make any significant inroads in this regard. After the brutal killing of the father of the nation, Sheikh Mujibur Rahman, by anti-liberation forces in 1975, the nation was under direct or indirect military rule for 15 years. The state started the process of Islamization, and certain JI leaders and war criminals were given a second chance in the political system. The Liberation War and the severity of the atrocities committed by individual JI leaders and the party as a whole were rarely talked about during this period.

In the 1990s, Jahanara Imam, an author and mother of the esteemed guerilla freedom fighter, Rumi, established the Ghatok Dalal Nirmul Committee, which initiated a movement aimed at exposing war criminals. This movement received considerable attention and educated the younger generation about the atrocities committed by war criminals. However, after the passing of Jahanara Imam, the momentum of this movement slowed down. Since 2006, a growing number of bloggers and online activists have been writing about the Liberation War, with a particular focus on the role played by the JI party throughout the war and the post-Liberation period. These individuals have invested their attention, efforts, and skills toward an online movement. In this regard, they have not hesitated to use aggressive tactics and vulgar language to engage in an online battle against the JI and its student wing, the Islami Chhatra Shibir (Haq, 2018).

In 2008, the Awami League (AL) party promised to try war criminals in its electoral manifesto, which gained significant support from both the younger generation, including bloggers and online activists, as well as typical youths. As a result, when the AL-led government established the International Crimes Tribunal in 2010 to investigate and prosecute war criminals, it was generally welcomed by the majority of the population (*Economist*, 2013). In 2013, the tribunal issued its second verdict, sentencing Abdul Quader Mollah, a leader of JI, the largest Islamist political party in Bangladesh, to life imprisonment for five charges, including murder, torture, and rape. However, this sentence was deemed inadequate by online activists and supporters of the Liberation War who expressed their disappointment and frustration over social media platforms. Many people suspected that the government had an arrangement with the war criminals to secure JI's support in the upcoming election and subsequently influenced the tribunal to deliver a lenient sentence (Haq, 2018). As a result, bloggers and online activists who disagreed with the verdict urged the public to gather at Shahbag, the square of activism and the cultural center of Dhaka. Tens of thousands of people gathered at Shahbag demanding the death penalty for Mollah. The gathering continued for 17 days and resulted in protests across the country and abroad, where a sizable number of Bangladeshis reside. Subsequently, the protesters extended their demands to include the banning of JI and a boycott of institutions affiliated with the party, which opposed the country's independence in 1971.

In response to the Shahbag movement, the Bangladeshi parliament amended the International Crimes Tribunal Act of 1973, which allowed the prosecution to request a change in Abdul Quader Mollah's sentence from life imprisonment to the death penalty. In September 2013, the Supreme Court modified the sentence, and Mollah was hanged that December. The revision of the International Crimes Tribunal Act and Mollah's death sentence are considered to be indications of the Shahbag movement's success (Roy, 2020). The unanticipated Shahbag movement is considered a turning point in Bangladeshi society, offering a new opportunity for citizens to voice their dissent against political trends and traditional political parties (Shahaduzzaman, 2014). However, Zaman (2016) argues that the movement not only legitimizes intolerance but also exacerbates polarization regarding identity issues. It is also said that the movement created a division among the citizens, leading to violence and vulnerability in society and democracy. The protest also led to the rise of terrorist attacks, mob rule, and attacks against religious minorities, which will impact the shaping of democratic institutions and practices (Chowdhury & Abid, 2019). The role of traditional mainstream media in this regard is also found to be significant (Parvez, 2022). On the other hand, Roy (2019) suggests that the Shahbag movement prompted mainstream media to accept social media as a legitimate source of news. The author also argues that political activists and leaders have become interested in using social media to further their political agendas after the movement.

New Social Movements and Mobile Communication

Social movements have taken on diverse forms throughout human history. In contemporary times, new social movements (NSMs) have emerged, supplanting traditional movements. NSMs primarily center on human rights issues, such as environmental and women's rights, as opposed to materialistic qualities such as economic wellbeing that traditional movements focus on. NSMs emphasize social changes in identity, lifestyle, and culture, rather than pushing for specific changes in public policy. The "new middle class," as opposed to the lower classes in traditional movements, are the key actors in NSMs (Pichardo, 1997, p. 418). NSMs are characterized by social networks of supporters, rather than members, unlike pressure groups that have formal structures and members (Byrne, 1997). Some scholars regard NSMs as relatively disorganized. Media plays a crucial role in NSM tactics, including protest campaigns that aid in achieving their objectives (Nash, 2010). NSMs differ from the "old" labor movement with a high degree of tolerance for political and ideological differences, making them more appealing to broader sections of the population due to their "local- and issue-centered" character (Dorsey & Collier, 2018, p. 310). This does not necessarily require a highly agreed-upon ideology or agreement on ultimate ends. The group of people who are marginalized in terms of the labor market – such as students, housewives, and the unemployed – participate in collective actions resulting from their disposable resource of time, their position on the receiving end of social movements, and their high levels of education and access to information and resources that lead to questions of how society is valued (Charles, 2002).

Along with mobile phones, mobile media refers to wireless Internet, digital cameras, MP3 players, pagers, mobile game consoles, global positioning, and satellite navigation systems (Siapera, 2012). The portability feature of mobile media facilitates mediated social connectivity, overcoming the limitations of fixed devices. Besides voice calls and text messages, the interaction may occur through photos, videos, and other tools. Campbell (2013) argues that in addition to direct human interaction mobile communication "support other forms of media consumption, information exchange, and even tracing the movement of objects that are in motion." (p. 9). Social media used by portable devices can also be considered mobile media. According to Manuel Castells (2007), mobile phones constitute a potent means of fostering political autonomy through the use of independent channels of communication between individuals and groups. These devices form the basis of a new public sphere, which is characterized by a distributed network of mobile communication. This network facilitates personalization, interactivity, and high-volume communication, while allowing consumers to circumvent conventional media channels. Castells emphasizes that this decentralized structure of mobile communication enables the creation of autonomous forms of political engagement, forms that are not subject to the control of traditional power structures. Scholars also argue that mobile devices offer a practical alternative for the general public in developing nations where limited access, poor literacy, and high costs pose impediments to the adoption of computing devices (such as personal computers, laptops, and tablets) (Latham, 2007; Skuse & Cousins, 2008).

Rheingold (2002, p. 174) used the term "smart mob" or "intelligent swarm" to describe this type of organizing. He argues that mobile phones promote a horizontal, network-style organization that is defined by the autonomy of the "subunits" that make up the network and the absence of a central authority. Such "nonlinear causation of peers influencing peers" is what propels these intelligent crowds (Rheingold, 2002, p. 178). The mobile network may exist and take shape without being directed by a centralized authority because users of mobile phones maintain their autonomy while also influencing the rest of the network to which they are linked. So, the political ramifications of mobile phones in social movements are seen as offering an alternative, direct, and more egalitarian form of political organizing than previous mediated forms. This articulation of mobility and politics has also benefited from the emergence of digital cameras, which are frequently integrated into mobile phones. Witnessing is the most well-known political application of digital cameras. People have the chance to participate in a political event that is captured on video, uploaded online, and then forwarded to major broadcast media. More broadly, ICTs are viewed as alternative public spheres that may transcend the constraints of time and space as well as the constrictive state and corporate mainstream media (Kahn & Kellner, 2004).

To protest the North American Free Trade Agreement, the Internet was first used in the Zapatista of Chiapas movement in Mexico in the early 1990s (Castells, 1996). Successful forms of mobilization in this protest inspired other global social movements of that decade to use the Internet and mobile phones to share information, plan direct action, and coordinate efforts to address the economic and social injustices brought on by global capitalism (Meikle, 2002; Bennet, 2003). However, the most famous early use of mobile phones in a protest was done in the Philippines where text messages were widely used to organize and coordinate the protests that ousted then the Philippines President Joseph Estrada (Castells, 2007). Although almost 40 percent of people of the country were living on a $1 daily income at the time of the protests in 2001, around 100 million SMS messages were sent each day during the protests in the country (Bociurkiw, 2001). Mobile phones were also widely used in other protests such as the Seattle World Trade Organization protests in 1999, Gleneagles for the G8 summit in 2005, and the Athens riots in 2008. The utilization of mobile media in social movements has been revolutionized with the advent of Internet-enabled mobile phones and social media. This has been made evident by all recent movements (i.e., Arab Spring and Occupy Movements), which have inspired protesters and activists across the world to utilize this new technology to mobilize individuals (Eltantawy & Wiest, 2011).

Against this backdrop, this study inquires as follows: How is mobile communication used for the inception, mobilization, and organization of the Shahbag movement? How does mobile communication contribute to turning this online movement into an offline one?

Methodology

For data collection, this qualitative study applies three methods: covert observation, document analysis, and semi-structured in-depth interviews. First, the researcher was personally involved in the online and offline movement as a participant observer. To understand the online movement, the researcher used his social media accounts and collected relevant posts shared during the movement from February 5, 2013 to February 21, 2013. Second, to compare initial research findings, 35 newspaper articles and columns, written by analysts and the organizers and participants of the movement, published during the period, were collected through Google search. Finally, 30 interviewees were selected for the semi-structured interview. To ensure the representation of all stakeholders and locations of the protest, interviewees were selected through the purposive sampling method. Aged between 21 and 55 (M=28.23) years old, the respondents included organizers, participants, and journalists. Professions of the organizers and participants included activists, university students, entrepreneurs, teachers, bankers, doctors, journalists, social workers, and party activists. Interviewees' social media profiles, movement websites, social media groups, pages, and channels were also reviewed to supplement the interview data. Among the interviews, 25 were conducted in Dhaka and Chattogram. Five interviewees, who lived abroad and were involved in online activism and organizing protest programs outside of the country, were interviewed online. All interviews were conducted between July and November 2015, lasted from half an hour to more than one hour, and were audiotaped.

All but one of the interviewees consented to be identified in this study. All audiotaped interviews were carefully listened to and transcribed and translated into English to analyze using the qualitative content analysis technique. Each interview is analyzed as a separate unit in this study. The process of forming a collection of thoughts that are very near to one other is referred to as the "meaning unit." "Words, phrases, or paragraphs having elements connected to each other by their content and context are considered meaning units" (Graneheim & Lundman, 2004, p. 106). Codes and categories were created by compressing and abstracting the content. The theme was then created using codes and categories. Direct quotations were used to support the themes.

Findings

Various communication tools were used in the Shahbag movement to make it a success. Along with traditional media such as posters, banners, leaflets, booklets, pictures, arts, sculpture, songs, and poems, interpersonal communication tools such as rallies, meetings, slogans, public speeches, and plays were also used. Mainstream media – such as newspapers, radio, and television – promoted the movement and motivated general people, who do not have access to the Internet, to join it (Sharma, 2020). In addition, mobile media, such as mobile phones, social media, blogs, digital cameras, and wireless Internet, are used in the following ways.

Mobile Communication in Initiating the Movement

The tribunal pronounced the verdict of JI leader Abdul Quader Mollah at around 10:30 am on February 5, 2013. Besides mainstream media, the news of the life imprisonment sentence of the war criminal reached many people through social media. By using smartphones, many frustrated activists expressed their frustrations online. Some of them wanted to do something to protest the verdict immediately. In a Facebook status, Anha F. Khan, an activist, asked at around 2 pm, tagging nine other activists of Shahbag-based activism: "Can't a protest rally be organized at the Shahbag intersection today as a quick reaction?" (Khan, 2013).

However, one and a half hours earlier Nirjhar Mazumder in a Facebook post at around 12:30 pm declared a protest rally in Chattogram and urged all to join:

> We will gather and organize a protest rally at 3:30 pm in front of Chittagong Press Club. Next program will be declared from there. We would declare *hartal*[1] like tough program if the gathering becomes big.
>
> (Mazumder, 2013)

This Facebook post got 324 likes, 93 shares, and 71 comments. Many people supported him and expressed their willingness to join the protest. Rakibul Bashar Rakib, an online activist, wanted to take responsibility for the Shahbag gathering at Dhaka. Some of them, who reside in Dhaka, asked Nirjhar what they could do in Dhaka. Nirjhar provided them with mobile phone numbers of Badhan Shopnokothok and Rakib and requested Dhaka's residents to contact them. About the beginning of the movement, Nirjhar said,

> After hearing the verdict at around 10:30 am, I contacted Badhan bhai in Dhaka over mobile phone and declared a protest rally in Chattogram through Facebook. Later Imran bhai phoned me and following the conversations a Facebook event was created on behalf of BOAN by Badhan bhai and Rakib. Popular blogger Sedative Hypnotics took the responsibility to promote the event on social media. We mainly used our smartphones to do all of these.
>
> (N. Mazumder, personal communication, September 8, 2015)

The Facebook event, titled "Grand rally protesting the farce verdict of Quader Mollah by the tribunal," was created by the Blogger and Online Activist Network (BOAN), urging all to gather at Shahbag at 3:30 pm. The phone numbers of Badhan Shopnokothok and Imran H Sarker were also provided with the details of the event. Along with BOAN members, many Facebook users shared the event on their own Facebook walls, inviting all to join. For instance, a blogger shared this event on Somewhereinblog by saying: "Come to Shahbag intersection today at 3:30 pm to protest" (Aziz, 2013).

The convener of BOAN, Imran H. Sarker, joined the protest from the beginning and emerged as the spokesperson of the movement on February 8. He described the beginning:

> The verdict of Mollah disappointed us. But we knew that we had to protest it on the streets instead of in the virtual world. That's why I contacted the members of BOAN and decided to gather at Shahbag at 3:30 pm to protest the verdict. A Facebook event was created from the BOAN page. Hundreds of bloggers and online activists gathered at Shahbag at around 3:30 pm. Continuous online campaign motivated people to join the protest. At around 5:30 pm we decided to block the Shahbag intersection, and we found thousands of people had joined us by 8:00 pm.
>
> (I. H. Sarker, personal communication, July 15, 2015)

Towhidul Alam is one of the organizers who joined the protest rally with his wife from the beginning. He contacted others over the mobile phone:

> After hearing the verdict, I discussed it with Nirjhar over the phone and requested him to organize a protest rally in front of Chittagong Press Club. Then I phoned Badhan and told him to create a Facebook event. He was in Imran bhai's house at that time. Later, he created the event, and I shared it on Facebook. When we reached Shahbag at around 3:40 pm, many online and Shahbag-centric activists gathered there by this time.
>
> (T. Alam, personal communication, July 10, 2015).

The results show that mobile communication, made through mobile phones, social media, blogs, and wireless Internet, played a crucial role in initiating the Shahbag movement. Firstly, many people got the news of Mollah's verdict and expressed their frustrations on social media used by mobile phones and wireless Internet. Secondly, besides communicating with each other, organizers used mobile communication to organize protest rallies and to reach supporters of the movement quickly. Finally, supporters got the news of protest rallies through mobile communication and joined the offline movement at Dhaka and Chattogram.

Mobile Communication in Mobilization

After gathering at Shahbag and Chittagong Press Club, many protesters captured photos and videos with digital cameras and immediately shared them on social networking sites by mobile Internet, inviting people to join them. In the evening, activist Parvez Alam, an organizer of the movement, wrote on Facebook from Shahbag:

> Those who are yet to join Shahbag will miss a great event. A huge number of people have gathered here ... Songs and slogans are blended here. Come here. Occupy Shahbag. Until we achieve the death sentence (for Mollah).
>
> (Alam, 2013)

Getting inspiration from this type of posts, many activists and social media users moved out toward the protest locations. Many of them shared their decisions to join the movement on social media on their way and inspired others to join. Sheikh Rokon, a journalist and an activist, wrote on Facebook: "Going to Shahbag ... What about you?" (Rokon, 2013).

Although many journalists had gathered at Shahbag by the evening, it was not telecast live by any TV channels until 10 pm. Besides sharing posts on social media, organizers requested television channels to telecast the rally live so that non-social media users would get the motivation to join the movement. Regarding this period, Shariful Hasan, an online activist and journalist, said,

> When many people joined the protest at Shahbag, I requested my colleagues in different TV channels over the phone to live telecast the protest. Finally, Channel 24 started to telecast the protest live at around 10:00 pm and general people across the country got the opportunity to watch what was happening at Shahbag.
>
> (S. Hasan, personal communication, July 22, 2015)

Live coverage of the protest by TV channels and online media motivated a broad spectrum of people to join the protest and spread the movement across the country. Taking inspiration from the Shahbag and Chattogram, supporters of the movement organized similar rallies in many cities across the country. Mainstream media published the news of these gatherings along with the Shahbag protest. The organizers of those protests created Facebook groups, pages, and events to communicate with people. Members of these groups regularly shared protest-related posts, pictures, cartoons, and news links from the protest venue through mobile devices to motivate other people to join.

Among the protests across the country, the protest in Sylhet was one of the biggest. Getting inspiration from Shahbag: hundreds of people gathered at Shahid Minar premises in the city and stayed for several consecutive days. Kabir Ahmad, an activist and an organizer of Sylhet gathering, describes this:

> When we came to know that people were gathering at Shahbag to protest the verdict, we decided to gather at Shahid Minar. Following the decisions, we created a Facebook event and invited all to join us. To reach the people who didn't only use Facebook, we contacted many people over the phone and through text messages and invited all to gather at Shahid Minar. Many people informed others about the program themselves. I also got some SMS from others. Finally, we found that along with us, many general people joined the protest program and took positions at the Shahid Minar for several consecutive days for the first time in Sylhet.
>
> (Ahmad, 2015)

The protest was not confined only to the country's territory but spread across the world, where the Bangladeshi people reside. Like the protesters in the country, mobile media, especially the mobile phones, social media, and messaging apps, helped the expatriates to mobilize people and other resources to organize solidarity programs with the movement. An organizer from Sweden said this:

> I knew about the protest from Facebook posts and was motivated to do something. After discussions with friends over phones and Facebook messages, we decided to gather at Stockholm University to express our solidarity with the movement. We invited our friends, through Facebook posts and over phones. We also created a Facebook group to share protest-related news and to motivate people in Sweden.
>
> (Anonymous, personal communication, September 20, 2015)

Along with political parties, many non-partisan people who often avoid political programs also joined the protest. Mobile communication played an important role in motivating them. Manjur Hossain, a banker, said:

> Though I was frustrated with the verdict of Mollah, I got the motivation to join the movement from Facebook posts. I found that the movement was not the same as other protests. I saw on Facebook that many of my friends joined the protest, which also inspired me.
>
> (M. Hossain, personal communication, August 18, 2015)

Many non-partisan people joined the protest as they found a moral ground there over the personal interest of an individual or social group. Sumon Mostofa, a protester, said,

> I joined the movement from my moral point of view. Social media posts made me believe that we have no moral right to try any criminal until we can bring war criminals to trial.
>
> (S. Mostofa, personal communication, September 18, 2015)

The present study finds that both mobile and traditional communication methods were utilized in the movement to mobilize individuals, with mobile media playing a pivotal role. While many mobile media users expressed on social media their dissatisfaction with the verdict, they were unsure whether to take any action in the physical world. However, when they discovered through mobile media that their peers, acquaintances, and social media influencers[2] were participating in the movement and inviting them to join, they became motivated to participate in the physical protest. The sharing of photos and videos on mobile media created a perception that an extraordinary event was taking place, and it was their moral responsibility to partake in the movement. Additionally, mobile media were employed to encourage television channels to broadcast the event live, which increased the credibility of the protest and motivated both Internet and non-Internet users to participate in the movement.

Shaping the Organizational Structure

Many individuals who joined the protests physically, along with those who supported the cause, shared movement-related information online independently and invited people to join the movement. Describing the role of mainstream media and the importance of mobile media in the protest, Akramul Hoque Samim, a blogger and organizer, said,

> Along with mainstream media, social media helped us a lot in organizing the protest. We regularly shared movement related updated information on our websites, blogs, and social media groups and pages by using mobile phones and computers. Many internet users asked us many questions about the protest and we answered them online.
>
> (A. H. Samim, personal communication, September 17, 2015)

One of the important advantages of online campaigns is the extensive and instantaneous reach to an online audience. Due to the interactive characteristic, social media users can acquire a sense of belonging to an imagined community. Mahbub Rashid, an activist, who joined the movement both online and offline, said,

> Those who could not participate with us physically were virtually involved with us through Facebook and mobile phones. Social media was also helpful for us

> in understanding public sentiment. Apart from this, we used mobile phone to communicate with each other and to coordinate the protest.
>
> (M. Rashid, personal communication, November 20, 2015)

It is found that the people independently organized protest rallies, across the country and abroad, without making any direct contact with anyone at the Shahbag protest. However, they followed the updates of the Shahbag protest on Facebook and observed the program declared from there.

The organizational structure of the Shahbag movement was a new experience for the activists of the country. Traditionally, people who join a rally, called by a political party or organization, know each other and have a strong connection with the central body. Activist Tapu Raihan analyzes this:

> We found many unknown non-partisan people joined in the Shahbag movement. However, they did not need to follow any direction made by any central body but rather engaged in the protest independently. Social media played an important role in inspiring them to join the protest since they could easily know about the movement's goals and objectives. The interactive tool helped us a lot to reach a huge number of people across the world and to win public support for the cause.
>
> (T. Raihan, personal communication, September 20, 2015)

Mobile media have brought a dramatic change in the recent social movements in Bangladesh. The traditional organizational pattern has been changed, and the Shahbag movement is an excellent example to understand this trend. Omar Shehab, a columnist, wrote in a newspaper,

> One of the main characteristics of the Shahbag protest was that it was universal. There was no organizational structure in the beginning … Shahbag Protest was a mass protest. Nobody came there to become familiar; rather, they came in response to their emotion (for the country).
>
> (Shehab, 2014)

The findings show that the organizational structure of the Shahbag movement was informal, loose, and flexible. Consistent with existing literature on new social movements, mobile communication was instrumental in shaping the movement's unconventional organizational structure, which was unique in the country's history of social movements. The movement served as an example of how mobile phones and social networking sites like blogs and Facebook have brought a new dimension to social movements in countries with low socio-economic and ICT status.

Discussion and Conclusion

This study aims to explore the role of mobile communication in the inception, mobilization and organization of the Shahbag movement and how it contributed to the transformation of the online movement into an offline one. The findings of this study are consistent with previous research (i.e., Haq, 2018) and suggest that the Shahbag movement was not initiated suddenly by young bloggers and activists, but rather was built on a long history of online and offline activism regarding the demand for the trial and capital punishment of war criminals. The sentence of life imprisonment for a war criminal frustrated the online activists who had been demanding trial and punishment for years. Furthermore, the perceived entente between the government and war criminals made them angry. This frustration and anger led to offline rallies that were not just demanding justice for one war criminal but were also a space for expressing frustrations and dissatisfaction with the trial process and upcoming verdicts. Furthermore, due to the small size of the online community and the relative anonymity of the participants, the individuals involved in the movement may have sought to create a visible presence. This sentiment was shared by many pro-liberation and non-political individuals who saw the protests as an opportunity to express their frustration without attending a political party's program. Consequently, the online movement transformed into an offline one, overcoming challenges related to limited access to ICTs. Consistent with the findings of previous research on contemporary social movements (Eltantawy & Wiest, 2011), this study reveals that mobile communication played a significant role in initiating the movement by allowing organizers and supporters to quickly exchange information. Additionally, the study demonstrates that mobile communication was instrumental in mobilizing individuals to participate in the movement by showcasing the involvement of friends, acquaintances, and influential figures. Although the protests are commonly referred to as the "Shahbag protest" (Chowdhury & Abid, 2019), this study highlights that the first rally was actually announced in Chattogram, and that mobile communication facilitated the organization of simultaneous rallies in two different cities.

The initial lack of coverage of the protest from mainstream media, due to the relatively unknown status of the bloggers and online activists, led organizers of the Shahbag movement to rely heavily on mobile communication, specifically social media, to mobilize participants in its early stages. Like many other recent movements (i.e., Arab Spring and Occupy Movement), through real-time communication, the movement was able to gain momentum and increase in size until it became newsworthy for mainstream TV channels. As of 2013, when only 6.7 percent of people in Bangladesh had access to the Internet, TV served as the primary source of breaking news, and social media was generally considered unreliable for news. The subsequent coverage of the protest on TV not only informed the public but also inspired frustrated pro-liberation individuals throughout the country to join the movement. While mobile communication played a significant role in initiating the Shahbag movement, it would not have been successful in achieving its goals without the use of traditional communication tools. This highlights the importance of a combination of both traditional and modern communication tools in mobilization efforts.

The Shahbag movement shares common characteristics with the Arab Spring, a series of uprisings in the Middle East and North Africa in 2010. Both movements were self-organized and autonomous. Like the Arab Spring, the Shahbag movement spontaneously claimed public spaces for protests and demonstrations, extending the limits of parliamentary democracy to allow citizens to express dissent, differences, and antagonism in the streets and engage directly with political actors for transformation and deliberation. Furthermore, the non-traditional nature of the Shahbag movement drew more people into it, as it was not a party-centric movement. However, despite the absence of traditional organizational control, the Shahbag movement had specific demands that motivated people to participate in the protest. The organizers used mobile communication to make their objectives clear to people and overcome time and space limitations, making the movement short-term goal oriented. Based on these findings, it is apparent that mobile communication can make a goal-oriented movement successful in a country with low socio-economic and ICTs status if citizens and mainstream media support the cause.

This study highlights that the organizational structure of the Shahbag movement, consistent with the new social movement theories, was initially informal, loose, and flexible due to its dependence on mobile communication. Mobile media enabled individuals to independently invite unknown people, disseminate information, and create protest materials, leading to spontaneous participation in the movement. Moreover, mobile communication, which constitutes a new type of network in Bangladesh society, not only played a vital role in initiating and mobilizing the movement but also in determining the relationship between organizers and participants. Like the organizing characteristics of Rheingold's (2002, p. 174) "smart mob" or "intelligent swarm," there was no hierarchy, but rather a horizontal relationship between organizers and supporters in the Shahbag movement. This new form of relationship, developed through mobile communication, made the movement an ideal example of a new social movement where people mainly participated on moral grounds. The study argues that the combination of both traditional mobilization tools and mobile communication strengthens mobilization actions, defines the organizational structure, and gives mobile communication new significance in Bangladesh society.

The main limitation of the study is the difficulty in selecting interviewees, as most of the organizers and participants are "unknown." Some of the interviews were taken through emails and Facebook messages. It would be better if I could have taken all interviews face-to-face. However, the study finds consistent results with existing studies and directs to further research. The study suggests that future researchers can analyze songs, poems, and slogans produced and distributed through mobile communication during the protest to better understand the movement's characteristics and how it contributed to increasing intolerance and to developing polarization in Bangladesh society.

Notes

1 Hartal refers to a widespread demonstration where various establishments such as workplaces, offices, shops, and courts are completely shut down. It shares similarities with a labor strike and serves as a form of civil disobedience. Apart from being a general

strike, it also entails the voluntary closure of schools and businesses. This approach is used to appeal to the government and urge it to overturn a decision that is deemed unpopular or unacceptable.

2 Social media influencers are individuals who have established credibility and expertise in a particular area. They frequently publish content related to that area on their preferred social media platforms and attract significant numbers of dedicated and engaged followers who closely follow their opinions.

References

Ahmad, K. (2015, February 5). *Gonojagoron: Sylhet Prekshit. Sylhet Today*. Retrieved from www.sylhettoday24.com/opinion/details/8/44

Alam, P. (2013, February 5). *Those who are yet to join Shahbag will miss a great event. A huge number of people have gathered here... Songs and slogans are blended here. Come here. Occupy Shahbag.* [Facebook status update]. Retrieved from www.facebook.com/parvezalambd?fref=ts

Aziz, P. (2013, February 5). Kosai kaderer fansi chai [Blog post]. Retrieved from http://m.somewhereinblog.net/mobile/blog/paglaaziz/29761869

Bass, G. (2013, November 19). Looking Away from Genocide. *The New Yorker*. Retrieved from www.newyorker.com/news/news-desk/looking-away-from-genocide

BBS (2022). *Population & Housing Census 2022*. Bangladesh Bureau of Statistics. Retrieved from https://drive.google.com/file/d/1Vhn2t_PbEzo5-NDGBeoFJq4XCoSzOVKg/view

Bennett, W. L. (2003). New media power: the internet and global activism. In: N. Couldry and J. Curran eds. *Contesting Media Power: Alternative media in a networked world* (pp. 17–37). Rowman & Littlefield Publishers.

Bociurkiw, M. (2001, September 10). Revolution by Cell Phone. *Forbes*. Retrieved from www.forbes.com/asap/2001/0910/028.html

Business Standard (2020, November 4). Bangladesh's telecommunication industry fifth-largest in Asia-Pacific. *The Business Standard.* Retrieved from www.tbsnews.net/bangladesh/bangladeshs-telecommunication-industry-fifth-largest-asia-pacific-153529

Business Standard (2022, July 27). 37% adults in country are internet users. *The Business Standard.* Retrieved from www.tbsnews.net/tech/37-adults-country-are-internet-users-466238

Byrne, P. (1997). *Social Movements in Britain: Theory and practice in British politics*. Psychology Press.

Campbell, S. W. (2013). Mobile media and communication: A new field, or just a new journal? *Mobile Media & Communication, 1*(1), 8–13.

Castells, M. (1996). *The Rise of the Network Society – the Information Age: Economy, society and culture*. Blackwell.

Castells, M. (2007). Communication, power and counter-power in the network society. *International Journal of Communication, 1*(1), 29.

Castells, M., Fernandez-Ardevol, M., Qui, J.L. & Sey, A. (2007). *Mobile Communication and Society: A global perspective*. The MIT Press.

Charles, N. (2002). *Feminism, the State and Social Policy*. Macmillan.

Chowdhury, A. R., & Abid, A. (2019). Emergent protest publics in India and Bangladesh: a comparative study of anti-corruption and Shahbag protests. *Protest Publics: Toward a New Concept of Mass Civic Action*, 49–66.

Dorsey, A. & Collier, R. (2018). *Origins of Sociological Theory*. ED-Tech Press.

Economist (2013, September 17). Final sentence. *The Economist*, Retrieved from www.economist.com/banyan/2013/09/17/final-sentence

Eltantawy, N., & Wiest, J. B. (2011). The Arab spring| Social media in the Egyptian revolution: reconsidering resource mobilization theory. *International Journal of Communication*, *5*, 18.

Graneheim, U. H. & Lundman, B. (2004). Qualitative content analysis in nursing research: concepts, procedures and measures to achieve trustworthiness. *Nurse Education Today, 24*(2), 105–112.

Haq, F. (2018). Shahbag Movement: Extended Conflict between Secularism and Religiosity in Bangladesh? *Bangabidya: International Journal of Bangladesh Studies*, *10*, 328–338.

Harlow, S. (2012). Social media and social movements: Facebook and an online Guatemalan justice movement that moved offline. *New Media & Society*, 14 (2), p. 225–243.

Hasan, Z. (2014). *Shahbag Theke Hefazat.* Dhaka: Adarsha.

Internet World Stats (2022). Bangladesh. *Internet World Stats*. Retrieved from www.internetworldstats.com/asia.htm#bd

Kahn, R. & Douglas, K. (2004). New media and internet activism: from the "Battle of Seattle" to blogging. *New Media & Society, 6*(1), 87–95.

Katz, J. E. & Aakhus, M. (Eds.). (2002). *Perpetual Contact: Mobile communication, private talk, public performance*. Cambridge University Press.

Kemp, S. (2022, February 22). Digital 2022: Bangladesh. Data Reportal. Retrieved from https://datareportal.com/reports/digital-2022-bangladesh#:~:text=There%20were%2052.58%20million%20internet,at%20the%20start%20of%202022

Khan, A. F. (2013, February 5). *Can a protest rally be organized at Shahbag intersection today as a quick reaction?* [Facebook status update]. Retrieved from www.facebook.com/anha.f.khan?fref=ts

Latham, K. (2007). SMS, communication, and citizenship in China's information society. *Critical Asian Studies*, *39*(2), 295–314.

Majumder, N. (2013, February 5). *We will gather and organize a protest rally at 3:30pm in front of Chittagong Press Club. Next program will be declared from there. We would declare hartal like tough program, if the gathering would be big.* [Facebook status update]. Retrieved from www.facebook.com/nirjharmazumder?fref=ts

Meikle, G. (2002). *Future Active: Media activism and the Internet*. Routledge.

Nash, K. (2010). *Contemporary Political Sociology: Globalization, Politics and Power*. Wiley-Blackwell.

Parvez, S. (2022). Understanding the Shahbag and Hefajat movements in Bangladesh: a critical discourse analysis. *Journal of Asian and African Studies*, *57*(4), 841–855.

Pichardo, N. A. (1997). New Social Movements: A Critical Review. *Annual Review of Sociology*, *23*, 411–430.

Rahman, A. & Hasan, M. (2022). From local to global: networked activism against multinational extractivism. *Review of Communication, 22*(3), p. 231–255.

Rahman, M. S., Haque, M. E., & Afrad, M. S. I. (2020). Utility of mobile phone usage in agricultural information dissemination in Bangladesh. *East African Scholars Journal of Agriculture and Life Sciences*, *3*(6), 154–170.

Rheingold, H (2002). *Smart Mobs: The Next Social Revolution*. Perseus.

Rokon, S. (2013, February 5). *Going to Shahbag... What about you?* [Facebook status update]. Retrieved from www.facebook.com/SheikhRokon?fref=ts.

Roy, A. D. (2020). *Not All Springs End Winter*. Adarsha.

Roy, R. K. (2019). Online Activism, Social Movements and Mediated Politics in Contemporary Bangladesh. *Society and Culture in South Asia*, *5*(2), 193–215.

Shahaduzzaman (2014). *Shahbag 2013*. Agami Prakashani.

Sharma, S. (2015). Nirbachone Samayik Yogayog Madhyom: Bastobota o Somvabona [Social Media in Election: Reality and Prospects]. *The Chittagong University Journal of Social Sciences, 33*, 255–276.

Sharma, S. & Ahsan, R. (2015). (Mis)Representations of the Liberation War of Bangladesh in Bollywood Films. *The Chittagong University Journal of Social Sciences, 33*(1), 115–128.

Sharma, S. & Islam, A. (2022, July 11–15). *Using without Knowing? Social Media Literacy in the Context of Privacy and Surveillance* [Paper presentation]. IAMCR annual conference. Beijing, China.

Sharma, S. & Rahaman, A. (2018, July 18). *Citizen Journalism in Bangladesh: Journalism with Fear?* [Paper presentation]. First International Conference on Media, Communication and Journalism, University of Chittagong, Chittagong, Bangladesh.

Sharma, S. (2014). Environmental movements and social networking sites in Bangladesh. *International Journal of Innovation and Sustainable Development, 8*(4), 380–393.

Sharma, S. (2020). Unanticipated Movements: Examining the Role of ICT in Leadership and Decision Making during the Shahbag Protests in Bangladesh. *Asiascape: Digital Asia, 7*(3), 187–210.

Sharma, S. (2023). Understanding female *jihadism* in Bangladesh: New trends in "new normal"? *Journal of Asian and African Studies.* doi: 10.1177/00219096231176737

Sharma, S., Ahmed, S. & Hossain, M. M. (2022). Meaning of New Means: Exploring the Economic Impact of Mobile Banking on Rural Bangladesh. *International Journal of Innovation in the Digital Economy (IJIDE), 13* (1), 1–18.

Shehab, O. (2014, February 5). *Shahbag theke Gonojagoron Moncho* (Shahbag to Gonojagoron Moncho). Bdnews24.com. Retrieved from http://opinion.bdnews24.com/bangla/archives/14922

Siapera, E. (2012). *Understanding New Media.* SAGE

Skuse, A., & Cousins, T. (2008). Getting connected: The social dynamics of urban telecommunications access and use in Khayelitsha, Cape Town. *New Media & Society, 10*(1), 9–26.

Udupa, S., Venkatraman, S., & Khan, A. (2020). "Millennial India": Global digital politics in context. *Television & New Media, 21*(4), 343–359.

World Bank (2020). The World Bank in Bangladesh. Retrieved on January 26, 2021 from www.worldbank.org/en/country/bangladesh/overview#1

Zaman, F. (2016). Agencies of Social Movements: Experiences of Bangladesh's Shahbag Movement and Hefazat-e-Islam. *Journal of Asian and African Studies, 53*(3), 339–349.

Part IV

Mobile Communication in the Networked Society

11 Reclaiming Power on Social Media

A Social Network Analysis of the #VeryAsian Movement on Twitter

Lei Guo and Jeremy Harris Lipschultz

Introduction

The increase of anti-Asian hate speech, harassment, microaggression, and crimes has been increasing since the COVID-19 pandemic began (Gover et al., 2020). The Federal Bureau of Investigation (FBI) documented more than 9,000 anti-Asian hate incidents reported from March 2020 to June 2021 to Stop AAPI (Asian American Pacific Islander) (Findling et al., 2022).

Hate speech and violent content targeting Asian and Asian American communities are widely disseminated across potentially powerful social media channels. A recent study found a surge of as much as 900 percent in online abuse toward Asian and Asian American communities since the beginning of the pandemic (The Cybersmile Foundation, 2022). The spread of such online abuse has even placed differential burdens on all Asian Americans.

The networking feature of social media, however, also has enabled some pushback and collective organizing against racism. Social media may help affected Asian and Asian American communities to reclaim power and space. The hashtag "#VeryAsian" rapidly increased on Twitter at the beginning of 2022 after a viewer left a racist voice message for a KSDK Korean American news anchor, Michelle Li, who shared transitional New Year's food in the newscast and said that she ate dumpling soup in celebration of this special day. This viewer commented on this by saying, "She's being very Asian. She can keep her Korean to herself" (Bellamy-Walker, 2022, para. 4). Li launched the #VeryAsian hashtag after sharing the racist voicemail on Twitter. The video and the hashtag "Very Asian" became popular on Twitter, Instagram, and many other social media platforms. Supporters use the #VeryAsian hashtag in social media posts to thank Li for sharing her heritage and taking a stand against Asian hate. Such conversations on Twitter generate networks with identifiable structures as users reply to and mention each other in tweets.

In this context, Twitter appears to be the primary social media platform for an emerging social movement to fight against online hate. However, there is a limited literature focusing on the role Twitter plays in mediating social movements in the Asian/Asian American community. A real-time dynamic examination of the general public's opinions regarding the crisis on Twitter is helpful for developing effective strategies to manage the crisis and diminish hostility toward particular

DOI: 10.4324/9781003328896-15

groups. Therefore, to fill the research gap, by using the Social Network Analysis (SNA) method, the main purpose of the present exploratory study is to gain a greater understanding of the spread pattern of the related messages by examining the betweenness and centrality within Twitter discussing #VeryAsian from January through December 2022. The findings of this study will inform further intercultural communication research within a heated political environment.

The chapter is structured as follows: First, a literature review is conducted on the introduction of the #VeryAsian Movement on Twitter as a research context and on the theoretical framework of social networks and the use of social media by minority communities; second, the rationale and process of using SNA as the research method are introduced; third, findings about the diffusion patterns of #VeryAsian tweets over time and suggestions for future research are discussed.

Literature Review

The #VeryAsian Movement

Asian hate may be placed within a larger context of pandemic-driven rising polarized political communication across social media channels. Social media communication can be considered a "super-spreader" of misinformation, disinformation, and conspiracy theories (Johnson et al., 2022, p. 67). Misinformation is an "umbrella concept" that "refers to various forms of misleading information that are inadvertently false and are shared without the intention of spreading falsities or causing harm" (p. 68). Disinformation, however, is "deliberately false or misleading" (p. 71). While racial prejudices traditionally were spread by traditional media (Lipschultz & Hilt, 2003), the lack of gatekeeping allowed non-professionals to spread hate directly to mass audiences.

Former president Donald Trump's racist language on social media, such as referring to COVID-19 as the "Chinese Virus" and "Kung Flu," corresponded to an increase in anti-Asian hate speech and a rise in discrimination and xenophobic attacks on Asian and Asian American communities (Benjamine, 2021; Kim & Kesari, 2021). Such misinformation and disinformation has been identified by computer-mediated communication and social media as ongoing issues across social media sites (Lipschultz et al., 2022). Trolls, bullies, and those accounts using intimidation may spread hate speech about ethnic minorities (Page, 2022). Specifically, cyberbullying may include gaslighting, which is "a type of insidious, and sometimes covert, emotional abuse where the abuser makes the target question their judgments and reality" (Page, 2022, p. 380). In response to such online hate speech and cyberbullying targeting the Asian/Asian American community, a wave of counter-anti-Asian movements, such as "StopAsianHate," has been mobilized on social media platforms.

In this context, the hashtag #VeryAsian has trended on Twitter, Instagram, and many other social media platforms after a news viewer left a racist voicemail for a local television reporter, Michelle Li. As a Korean American, Li shared her own tradition while reporting a story about New Year's Day food, saying, "I ate dumpling soup. That's what a lot of Korean people do." She received a voicemail that complained that Li was "being very Asian" and she should "keep her Korean to herself." Li tweeted her video of listening to the voice message, which inspired a solidarity movement with thousands of people using the hashtag #VeryAsian to share their own traditions and heritage. Dozens of reporters, activists, and celebrities also tweeted, commented, and responded to each other to counter anti-Asian sentiment. With massive online and offline support, Li launched The Very Asian Foundation to amplify diverse AAPI voices and fundraise for other pro-Asian organizations. However, the existing literature about #VeryAsian and Asian hate is limited. Therefore, a real-time dynamic examination of social media users' opinions toward such a crisis is helpful to understand and manage the spread of information on social media.

The Use of Social Media by Minority Communities

With the features of public accessibility and near real-time feeds, social media has become a major platform to connect and enable conversations among almost all types and sizes of communities (Bahfen, 2018; Brough et al., 2020; Johnson & Callahan, 2013). The latest research conducted by Pew Research Center reported that 72 percent of Americans use social media sites. The percentage even rises to 84 percent when only the population between 18 and 29 is taken into account (Auxier & Anderson, 2021).

The existing literature about race and media use focuses more on understanding how ethnic minorities utilize new media and has revealed the main motivations for ethnic minorities to consume social media, including cultural adaptation, self-representation, online engagement, and community integration (Bahfen, 2018; Mittelstädt & Odag, 2015). However, there are differences by ethnicity in the use of social media platforms regarding the frequency of use, platform preference, and motivations. For instance, previous studies found that the photo-sharing platform is more prevalent among Hispanics than other minority communities, while black Americans are more likely to use social media for online political participation and social activism (Auxier, 2020; Krogstad, 2015).

As the fastest-growing ethnic group in the United States, Asians and Asian Americans, especially English-speaking Asian Americans, have accounted for a sizable portion of social media consumption (Budiman & Ruiz, 2021). In 2015, 95 percent of Asians and Asian Americans had access to the Internet, which is significantly higher than other minority groups' Internet accessibility (Perrin, 2016). Based on studies conducted by Pew Research Center and Nielsen, approximately 91 percent of Asian Americans own a smartphone, which increases their accessibility to the Internet and social media platforms, compared with 66 percent of whites, 62 percent of blacks, and 65 percent of Hispanics (Nielson, 2018; Perrin, 2016). Meanwhile, compared with the general population in the United States, Asians and Asian Americans are also reported to spend more time watching videos, using apps, and browsing websites on smartphones and become primary consumers on several leading social media platforms, such as Skype and Whatsapp

(Nielson, 2018). However, there has been limited research, specifically on social media use, among Asians and Asian Americans due to the relatively small share of the U.S. overall population and the diversity of origins within the group (Budiman & Ruiz, 2021; Charmaraman et al., 2018).

However, there is a long history of racism against Asian and Asian American communities in the United States. The latest focus group study conducted by Pew Research Center with 18 Asian origin groups in the United States. revealed that ignorance and misinformation about Asian and Asian American adolescents caused racial discrimination, differential treatment, and even dangerous situations in this community. Participants also acknowledged the typical stereotype of treating them as "forever foreigners" (Ruiz et al., 2022b). Employers in North America were also found to discriminate against applicants with Asian names, such as Indian, Pakistani, and Chinese names (Banerjee et al., 2018).

COVID-19 has also enabled the spread of anti-Asian racism and created national insecurity, and general xenophobia, which has dramatically increased the anti-Asian crimes since the pandemic began (Gover et al., 2020; Ruiz et al., 2022a). According to a national survey by AAPI Data, one in six Asian American adults experienced a hate crime in 2021 compared with one in eight in 2020 (Lee & Ramakrishnan, 2022). COVID-19-related racial discrimination was found to significantly contribute to the increased mental health distress and posttraumatic stress disorder symptoms of Asians and Asian Americans (Hahm et al., 2021; Park et al., 2021).

Social media platforms also account for spreading anti-Asian sentiment during the pandemic. The existing literature focuses on understanding how COVID-19 misinformation has driven the increasing amount of hate speech targeting Asian and Asian American communities. For example, with the anxiety and fear about the pandemic spreading on social media, Hohl et al. (2022) found that the volume of anti-Asian hateful tweets surged between November 2019 and May 2020. Exploratory research also revealed that frequent social media users were more likely to believe Asian and Asian Americans pose a realistic threat to America (Croucher et al., 2020). On the other hand, Yang et al. (2020) also confirmed that Asian and Asian Americans experienced more online discrimination during COVID-19 if they frequently engaged in social media private messaging, posting, and even browsing.

To counter hateful messages or support anti-racism activism, social media have also become an important tool in Asian and Asian American communities (Croucher et al., 2020). However, empirical attention to the pattern and effectiveness of Asian American social media use, especially during the social crisis, is largely overlooked. Therefore, further research to explore how Asians and Asian Americans navigate hateful messages and protect themselves on social media is needed.

The Social Networks on Social Media

The social network approach is a combination of the theoretical framework and analytical methods of networked relationships to analyze information flow

(Watanabe et al., 2021). This approach aims to account for individuals' communication behaviors by depicting the social network and interactional relationships among actors in the network (Emirbayer & Goodwin, 1994; Watanabe et al., 2021). From this perspective, the network mapping and statistical measures are utilized to assess each actor's position within the information diffusion process by considering density and centrality metrics, including degree, betweenness, and closeness (Knoke & Yang, 2019; Scott, 1988; Watanabe et al., 2021).

Social media like Facebook, Twitter, and Instagram, are constituted by the collection of websites and web-based systems that allow for synchronous and asynchronous mass interaction, conversation, and sharing among users of a network (del Fresno García et al., 2016; Hill et al., 2013). It is used by millions of users who collectively share interests, political views, and other activities to generate public conversations from near real-time interactions. This is to say that social media use is heavily dependent on networking, and such complex relationship structures that emerge on social media can be studied by using social network analysis (SNA) (del Fresno García et al., 2016).

The social network approach has been adopted in exploring social phenomena in many disciplines. Monaghan et al.'s (2017) systematic review study also found that in addition to social media, SNA is also increasingly used to map inter- and intra-organizational networks, as well as business networks. The latest studies have also utilized SNA to track online public conversations regarding social phenomena or online movements in recent years, such as #Metoo, #BlackLivesMatter, and #StopAsianHate movement. For instance, by analyzing 3,355 tweets, the authors revealed that social media users' posts were highly likely to become hateful after users were exposed to hateful content. However, counterspeech messages might discourage users from turning hateful (Ziems et al., 2020). Based on evidence from previous research, by mapping social media network spaces, we believe that the "natural" structures of the crowd and the conversation around the #VeryAsian movement can be represented using social network analysis.

Twitter as a Network of Public Opinion

As one of the most investigated topics in the mass communication field, public opinion academic research has been started since the early twentieth century (Heath et al., 2005). The earliest public opinion studies concentrated on local voters' attitudes toward election candidates and their voting intentions. Opinion polling or surveys are predominant techniques to measure public opinion (Heath et al., 2005; Kwon et al., 2016). However, because of the nature of self-report data, survey results can only partially represent public opinion.

Compared to opinion polling, social media is a space for users to express their opinions publicly. It provides a set of data on various topics and allows researchers to track them (Anstead & O'Loughlin, 2015; Stieglitz & Dang-Xuan, 2013). A growing body of literature focuses on using SNA to reflect social media users' opinions toward social and political issues. A recent study with 280 thousand COVID-19-related tweets in Ohio and Michigan demonstrated that the changes in

COVID-19-related tweet trends were related to local authorities' reactions and the number of daily confirmed positive cases.

Previous research also found significant distinctions between different forms of online activity and social media use that might influence users' opinions and political participation (Kahne & Bowyer, 2018). Among other social media platforms, Twitter stands out as a unique source of big data because of the near real-time nature of its content and the ease of accessing and searching its publicly available information (Ahmed, 2021; Ahmed et al., 2017). In addition, Twitter has also become a new media platform for social movements since users can connect with others, advocate for each other, engage in conversation, share resources, and support community members on a specific topic. For instance, research has examined how users tweet Black Lives Matter-related content to foster dialogue, engage the public, and promote action (Edringto & Lee, 2018).

The #VeryAsian movement emerged from a hashtag on Twitter and grew as an online and offline social movement in response to hateful sentiment toward Asian and Asian American communities on social media. It also plays an important role in the Stop Asian Hate protest. This movement, thus, offers the opportunity to examine the diffusion of opinions toward #VeryAsian that spread anti-Asian hate content and the responses to it. Therefore, the present study seeks to answer the following questions:

RQ 1: How do the #VeryAsian tweets show a spread of hate speech and the online community response to it?
RQ 2: How did the specific users contribute to the online interaction within their larger political communication context?
RQ 3: How do users describe their sentiment within this Twitter interaction?
RQ 4: To what extent did SNA data reflect expected concepts identified within studies about violence against Asians?

Research Design and Method

SNA has been recognized as a powerful tool for representing social network structures and information dissemination (del Fresno García et al., 2016). Social network communication typically happens within clusters of like-minded user accounts. Individuals express opinions within networks that reflect varying degrees of social ties. Using SNA, hashtags have been studied to comprehensively understand the spread of social movements on Twitter and other social media platforms, and to provide an overview of the intellectual structure of information diffusion (Himelboim et al., 2013).

Therefore, the present study explored a short-lived conversation on Twitter sparked by that New Year's Day 2021 #VeryAsian tweet and subsequent reaction. The data was collected from a Twitter search between January and December 2022. Users were included in the dataset if they entered a tweet during the time the data were retrieved or were mentioned or replied to in these tweets. The #VeryAsian hashtag was selected, as this was the most popular and in early January 2022 briefly became a trending topic on Twitter. Six network structures, namely polarized crowds, tight crowds, brand clusters, community clusters, broadcast networks, and support networks (Smith et al., 2014), were applied to the present study to examine the network structure of the #VeryAsian movement and interactions among users. According to Smith et al. (2004), polarized crowd and tight crowd are opposites of one another in terms of division; brand clusters and community clusters are different in terms of the density of clustered connections; and the broadcast network and support network feature the direction of the clustered connections.

NodeXL was used as a tool to analyze egocentric networks and betweenness and closeness centrality in the #VeryAsian movement on Twitter. Especially, egocentric network analysis allows for visualization of relational data that exist in tweeting #VeryAsian. Betweenness centrality measures the frequency at which an individual node lies on the shortest path connecting other nodes in the network. Finally, closeness centrality measures the average distance between an individual node and all other nodes in the network. These network shapes can consist of broadcast networks, polarized crowds, brand clusters, tight crowds, community clusters, and support networks.

Research Findings

RQ1 asks how the #VeryAsian tweets show a spread of hate speech and the online community response to it. According to SNA analysis, #VeryAsian Tweets had been widely distributed between January and April 2022 (See Figure 11.1). During the four-month data collection period, the #VeryAsian hashtag had been mentioned 23.35 thousand times and reached 206.76 million users on Twitter during our data collection period. Photos, gifs, and videos were the most used content type among analyzed tweets. As Figure 11.1 shows, during the first week of the #VeryAsian movement, 16,877 Tweets were generated by 11,962 users. However, the number has dramatically dropped to 168 Tweets in the first week of April 2022. As for Twitter users, among those who tweeted #VeryAsian-related content, 8.95 thousand were female users, while 6.1 thousand were males. The majority of users come from North America, followed by South Korea, the United Kingdom, Comoros, Sri Lanka, Australia, France, Japan, and China.

The SNA analysis, of 11,962 #VeryAsian-related Tweets between January 9 to 18, 2022, and 168 Tweets on April 8, 2022 shows follows, replies-to, and mentions relationships in tweets. The gray lines represent each "follow" relationship, and the red line for any "replies-to" and "mentions" relationship in a tweet. At the center of Group 1 is the account for Michelle Li, surrounded by a large number of users in Group 1 and other Groups who individually connect to Michelle Li's account. This density pattern is shared by many other groups that are identified by SNA. In contrast, only Group 3 is a community composed of densely connected users who all connect with each other and engage in conversations around the subject.

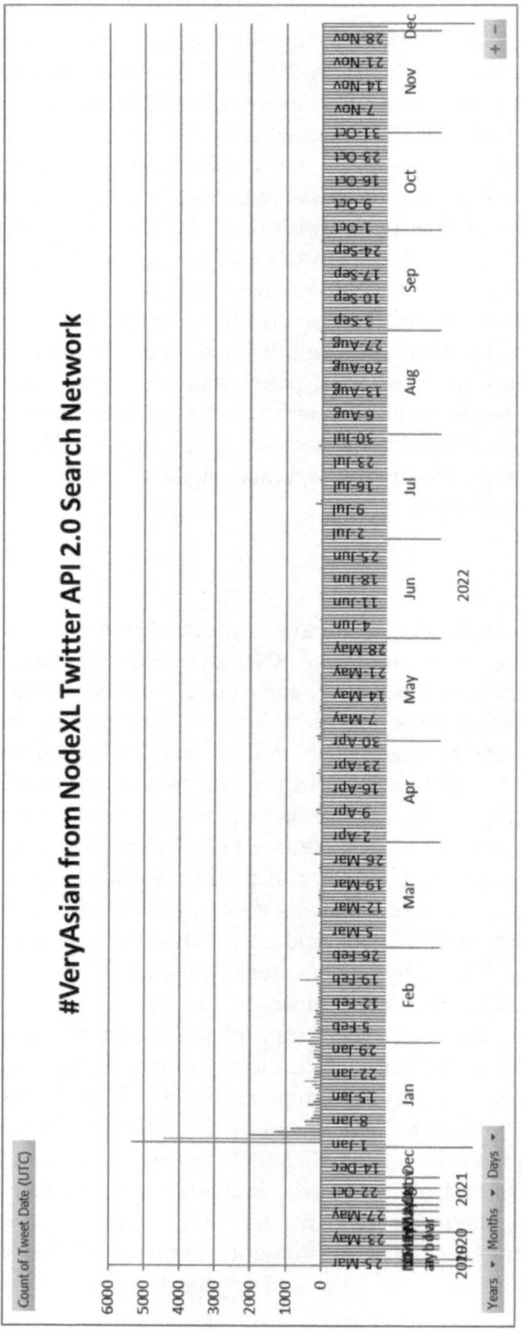

Figure 11.1 Number of Tweets Related to #VeryAsian in January and April 2022.

Source: Figure generated by NodeXL.

Figure 11.2 Top Prolific Users and Influencers in Tweeting #VeryAsian.

Source: Figure generated by NodeXL.

Later data from late March and early April was similar to the original findings. The initial January topic framing results were held, and there was not much in the way of a new iteration.

RQ2 asks how the specific users contribute to the online interaction within their larger political communication context. As shown from the social network maps, the majority of discussion groups of #VeryAsian were advocated by a few certain accounts. Further analysis revealed those top influencers are reporters and media professionals (See Figure 11.2).

RQ 3 asks how users describe their sentiment within this Twitter interaction. The sentiment analysis further classified 6.36 thousand tweets as having positive #VeryAsian sentiment, 10.96 thousand as having neutral sentiment, and 3.65 thousand as having negative sentiment. Overall, the number of positive and neutral tweets is almost five times more than negative tweets. The top five most positive sentiment words consisted of "proud," "like," "support," "love," and "celebrate," whereas negative sentiment words were "ashamed," "attacks," "limited," "racism," and "mad" (See Figure 11.3).

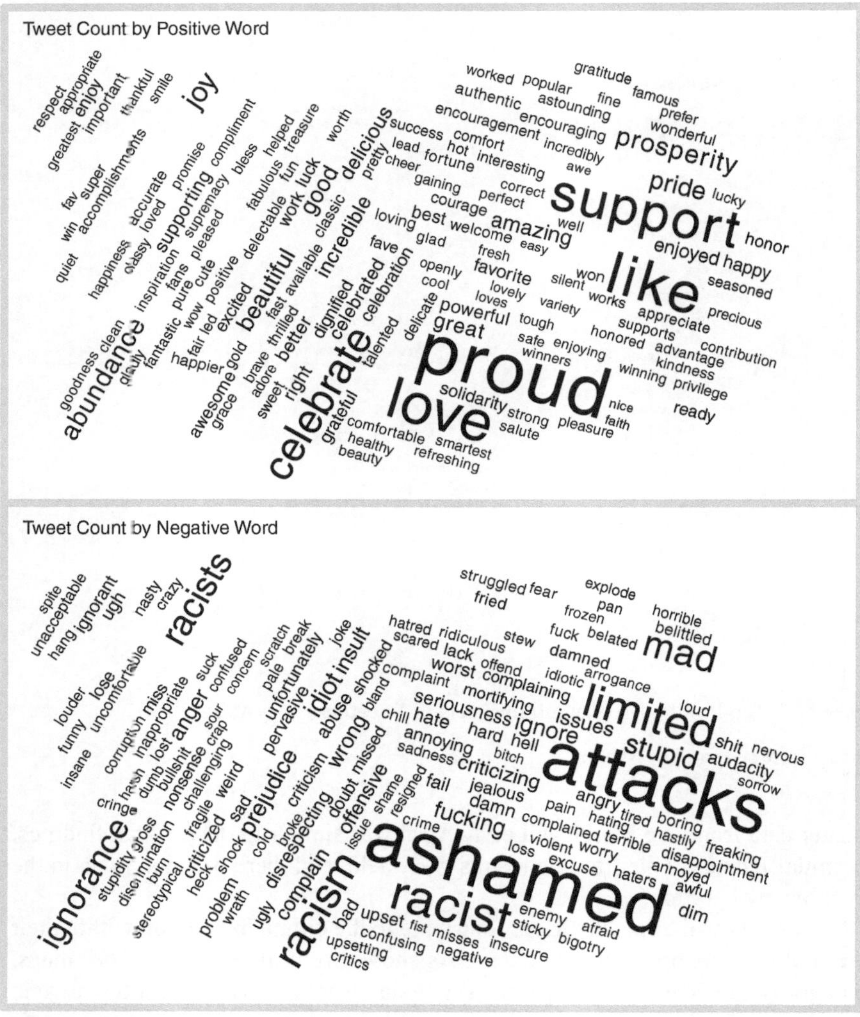

Figure 11.3 The Most Frequent Words Related to #VeryAsian.

Source: Figure generated by NodeXL.

In Figure 11.3, the "word clouds" show the most frequently used word stems across Twitter users' post descriptions related to #VeryAsian. The upper image is a word cloud formed from tweets of positive sentiment, while the lower image is formed from tweets of negative sentiment.

RQ 4 examines to what extent the SNA data reflect expected concepts identified within studies about violence against Asians. The topic modeling identified

Figure 11.4 Social Network Graphs of the Dominant Themes About #VeryAsian, with the Top Six Associated Words per the Theme.

Source: Figure generated by NodeXL.

six themes that dominated Twitter discussions of #VeryAsian, and those themes are labeled as VeryAsian in 2022, Dumplings, MichelleLiTV, Michelle, New Year, and StopAsianHate&VeryAsian. Figure 11.4 demonstrates these six themes, and each theme includes the six most frequently used words in tweets corresponding to that specific theme. The words are referred to as users when describing social network graphs, in which the node's size represents the frequency of a particular word appearing. "Love" and "T-Shirt" dominated the discussion of VeryAsian in 2022. "Mixed House" dominated the discussion of dumplings. In MichelleLiTV's discussion, "White," "Family," and "Love." "Food" and "Person" were frequently discussed in the theme of Michelle. "Love," "Eating," "Dim Sum," and "Happy New Year" dominated the discussion of New Year. "Long," "Diversity," and "White Women" dominated in "StopAsianHate & VeryAsian."

Discussion and Limitation

Social conversation on Twitter offers a wealth of social network data about controversial political issues and social movements (Hosterman et al., 2018; Murthy, 2012). Public conversations on social media, thus, are as important to document as other large public gatherings (Smith et al., 2014). Hateful speeches and sentiments toward Asian and Asian American communities have surged on social media since the COVID-19 pandemic began (Hohl et al., 2022). The #VeryAsian movement emerged in such a context and users from all over the world shared information on social media to celebrate their Asian culture and heritage. By analyzing the

#VeryAsian-related tweets between January and December 2022, the present study provides preliminary evidence of the social network structure of information flow of the #VeryAsian movement.

As with other online messaging movements (e.g., Edringto & Lee, 2018), the number of #VeryAsian-related tweets decreased gradually over time. In terms of demographic factors considered in our analysis, female Twitter users were more active than their male counterparts. There are also certain influential agenda setters and opinion leaders to advocate for other users, especially at the beginning of the movement. In line with previous research (Ziems et al., 2020), positive and neutral sentiments dominated the #VeryAsian tweets, which revealed that the use of public counterspeech messaging campaigns might be a potential solution against hateful sentiment on social media. However, the results also indicated that it is crucial for influential opinion leaders to post related information and increase the frequency of "replies-to" and "mentions" relationship in tweets to advocate the online social movement.

There are limitations that need to be acknowledged in this study as well. The present study only collected #VeryAsian-related tweets in four months. We found that the number of tweets dropped dramatically from January to April. However, users might generate different hashtags to be involved in this online public conversation. Future studies should continue to track #VeryAsian and similar Twitter hashtags, such as "very Asian," for evidence of development in what has been called social movements in the research (Edringto & Lee, 2018). In addition, even though our analysis finds that many Twitter users are actively engaging in the #VeryAsian movement and fighting back the hateful sentiments on social media, it is certainly the case that not all users will use social media in the same manner. Therefore, future studies of the #VeryAsian movement or related topics using SNA should also analyze the information flow on other social media platforms, as well as how special events could interactively impact networked behavior on different social media platforms. Finally, other research methods, like surveys, can be helpful in understanding users' certain networked behaviors on social media since SNA focuses more on describing relationships among social media accounts.

Conclusion

Since the COVID-19 pandemic, Asian and Asian Americans have been experiencing increasing discrimination (Yang et al., 2020). Social media, thus, has become an important tool for Asian and Asian Americans to copy with discrimination and/or support anti-racism activism (Croucher, 2020). The main purpose of this study is gaining a greater understanding of the information flow of #VeryAsian on Twitter and identifying the pattern of anti-hate speech messaging. By analyzing post, follows, replies-to, and mentions relationships in tweets, our research identified that media professionals and opinion leaders advocated #VeryAsian-related discussion in the early stage of this online campaign against hateful sentiment on social media. To continue to diminish hostility toward the Asian and Asian American communities, the results also showed that the continuous interactions between agenda setter and other users were needed to manage the public's opinions.

References

Ahmed, W. (2021, May 18). Using Twitter as a data source an overview of social media research tools (2021). Retrieved from *Impact of Social Sciences Blog*. https://blogs.lse.ac.uk/impactofsocialsciences/2021/05/18/using-twitter-as-a-data-source-an-overview-of-social-media-research-tools-2021/

Ahmed, W., Bath, P. A., & Demartini, G. (2017). Using Twitter as a data source: An overview of ethical, legal, and methodological challenges. *The Ethics of Online Research, 2*, 79–107. doi:10.1108/S2398-601820180000002004

Anstead, N., & O'Loughlin, B. (2015). Social media analysis and public opinion: The 2010 UK general election. *Journal of Computer-Mediated Communication, 20*(2), 204–220. doi:10.1111/jcc4.12102

Auxier, B. (2020, December 11). Social media continue to be important political outlets for Black Americans. Retrieved from *Pew Research Center* website. www.pewresearch.org/fact-tank/2020/12/11/social-media-continue-to-be-important-political-outlets-for-black-americans/

Auxier, B., & Anderson, M. (2021, April 7). Social media use in 2021. Retrieved from *Pew Research Center* website. www.pewresearch.org/internet/2021/04/07/social-media-use-in-2021/

Bahfen, N. (2018). The individual and the ummah: The use of social media by Muslim minority communities in Australia and the United States. *Journal of Muslim Minority Affairs, 38*(1), 119–131. doi:10.1080/13602004.2018.1434939

Banerjee, R., Reitz, J. G., & Oreopoulos, P. (2018). Do large employers treat racial minorities more fairly? An analysis of Canadian field experiment data. *Canadian Public Policy, 44*(1), 1–12.

Bellamy-Walker, T. (2022, Jan. 3rd). #VeryAsian hashtag goes viral after racist criticism of Korean American news anchor. Retrieved from *NBC News* website. www.nbcnews.com/news/asian-america/veryasian-hashtag-goes-viral-racist-criticism-korean-american-news-anc-rcna10777

Benjamin, E. (2021). Trump, the coronavirus pandemic, Asian American xenophobia, and humanistic psychology. *Journal of Humanistic Psychology, 61*(2), 244–259. doi: 10.1177/0022167820979650

Brough, M., Literat, I., & Ikin, A. (2020). "Good social media?": underrepresented youth perspectives on the ethical and equitable design of social media platforms. *Social Media+ Society, 6*(2), 1–11. doi:10.1177/2056305120928488

Budiman, A., & Ruiz, G. N. (2021). Asian Americans are the fastest growing racial or ethnic group in the U.S. Retrieved from *Pew Research Center* website. www.pewresearch.org/fact-tank/2021/04/09/asian-americans-are-the-fastest-growing-racial-or-ethnic-group-in-the-u-s/

Charmaraman, L., Chan, H. B., Chen, S., Richer, A., & Ramanudom, B. (2018). Asian American social media use: From cyber dependence and cyber harassment to saving face. *Asian American Journal of Psychology, 9*(1), 72–86. doi:10.1037/aap0000109

Croucher, S. M., Nguyen, T., & Rahmani, D. (2020). Prejudice toward Asian Americans in the COVID-19 pandemic: The effects of social media use in the United States. *Frontiers in Communication, 5*, 39. doi:10.3389/fcomm.2020.00039 del Fresno García, M., Daly, A. J., & Segado Sanchez-Cabezudo, S. (2016). Identifying the new influences in the Internet era: Social media and social network analysis. *Revista Española de Investigaciones Sociológicas, 153*, 23–40. doi:10.5477/cis/reis.153.23

Edrington, C. L., & Lee, N. (2018). Tweeting a social movement: Black Lives Matter and its use of Twitter to share information, build community, and promote action. *The Journal of Public Interest Communications, 2*(2), 289–289. doi:10.32473/jpic.v2.i2.p289

Emirbayer, M., & Goodwin, J. (1994). Network analysis, culture, and the problem of agency. *American Journal of Sociology, 99*(6), 1411–1454. doi:10.1086/230450

Findling, M., Blendon, J. R., Benson, J., Koh, H. (2022). Covid-19 has driven racism and violence against Asian Americans: Perspectives from 12 national polls. Retrieved from *Health Affairs Forefront* website. www.healthaffairs.org/do/10.1377/forefront.20220411.655787

Gover, A. R., Harper, S. B., & Langton, L. (2020). Anti-Asian hate crime during the COVID-19 pandemic: Exploring the reproduction of inequality. *American Journal of Criminal Justice, 45*(4), 647–667. doi:10.1007/s12103-020-09545-1

Hahm, H. C., Ha, Y., Scott, J. C., Wongchai, V., Chen, J. A., & Liu, C. H. (2021). Perceived COVID-19-related anti-Asian discrimination predicts post traumatic stress disorder symptoms among Asian and Asian American young adults. *Psychiatry Research, 303*, 114084. doi:10.1016/j.psychres.2021.114084

Heath, A., Fisher, S., & Smith, S. (2005). The globalization of public opinion research. *Annual Review of Political Science, 8*, 297–333. doi:10.1146/annurev.polisci.8.090203.103000

Hill, C. A., Dean, E., & Murphy, J. (2013). *Social media, sociality, and survey research.* John Wiley.

Himelboim, I., McCreery, S., & Smith, M. (2013). Birds of a feather Tweet together: Integrating network and content analyses to examine cross-ideology exposure on Twitter. *Journal of Computer-Mediated Communication, 18*(2),154–174. doi:10.1111/jcc4.12001

Hohl, A., Choi, M., Yellow Horse, A. J., Medina, R. M., Wan, N., & Wen, M. (2022). Spatial distribution of hateful Tweets against Asians and Asian Americans during the COVID-19 pandemic, November 2019 to May 2020. *American Journal of Public Health, 112*(4), 646–649. doi:10.2105/AJPH.2021.306653

Hosterman, A. R., Johnson, N. R., Stouffer, R., & Herring, S. (2018). Twitter, social support messages, and the# MeToo movement. *The Journal of Social Media in Society, 7*(2), 69–91.

Johnson, J. L., & Callahan, C. (2013). Minority cultures and social media: Magnifying Garifuna. *Journal of Intercultural Communication Research, 42*(4), 319–339. doi:10.1080/17475759.2013.842608

Johnson, T. J., Wallace, R., & Lee, T. (2022). How social media serve as a super-spreader of misinformation, disinformation, and conspiracy theories (pp. 67–84). In J. H. Lipschultz, K. Freberg and R. Luttrell (eds.), *The Emerald Handbook of Computer-mediated Communication and Social Media*. Emerald. doi: 10.1108/9781800715974.

Kahne, J., & Bowyer, B. (2018). The political significance of social media activity and social networks. *Political Communication, 35*(3), 470–493. doi:10.1080/10584609.2018.1426662

Kim, J. Y., & Kesari, A. (2021). Misinformation and hate speech: The case of anti-Asian hate speech during the COVID-19 pandemic. *Journal of Online Trust and Safety, 1*(1), 1–14. doi: 10.54501/jots.v1i1.13

Knoke, D., & Yang, S. (2019). *Social Network Analysis*. SAGE publications.

Krogstad, M. J. (2015, February 3). Social media preferences vary by race and ethnicity. Retrieved from *Pew Research Center* website. www.pewresearch.org/fact-tank/2015/02/03/social-media-preferences-vary-by-race-and-ethnicity/

Kwon, K. H., Bang, C. C., Egnoto, M., & Raghav Rao, H. (2016). Social media rumors as improvised public opinion: Semantic network analyses of twitter discourses during Korean saber rattling 2013. *Asian Journal of Communication, 26*(3), 201–222. doi:10.1080/01292986.2015.1130157

Lee, J., & Ramakrishnan, K. (2022, March 16). A year after Atlanta. As we honor the victims, we ask: How much has anti-Asian hate changed? Retrieved from *AAPI Data* website. http://aapidata.com/blog/year-after-atlanta/

Lipschultz, J. H., & Hilt, M. L. (2003). Race and local television news crime coverage. *Communication Faculty Publications 31*. https://digitalcommons.unomaha.edu/commfacpub/31/

Lipschultz, J. H., Freberg, K., & Luttrell, R. (2022). Introduction: Connecting CMC and social media research (pp. 3–19). In J. H. Lipschultz, K. Freberg and R. Luttrell (eds.), *The Emerald Handbook of Computer-mediated Communication and Social Media*. Emerald. doi: 10.1108/9781800715974

Mittelstädt, A., & Odag, Ö. (2015). Social media use and social integration of ethnic minorities in Germany: A new interdisciplinary framework. *Athens Journal of Mass Media and Communications, 2*(1), 21–32. doi:10.30958/ajmmc.2.1.2

Monaghan, S., Lavelle, J., & Gunnigle, P. (2017). Mapping networks: Exploring the utility of social network analysis in management research and practice. *Journal of Business Research, 76*, 136–144. doi:10.1016/j.jbusres.2017.03.020

Murthy, D. (2012). Towards a sociological understanding of social media: Theorizing Twitter. *Sociology, 46*(6), 1059–1073. doi:10.1177/0038038511422553

Nielsen (2018, May 8). Asian-American consumers are predictive adopters of new media platforms, online shopping and smartphone use. Retrieved from *Nielsen* website. www.nielsen.com/news-center/2018/asian-american-consumers-are-predictive-adopters-of-new-media-platforms/

Page, S. (2022). The transition to 24/7 trolls, bullies, and intimidation through social media. (pp. 377–395). In J. H. Lipschultz, K. Freberg and R. Luttrell (eds.), *The Emerald Handbook of Computer-mediated Communication and Social Media*. Bingley, UK: Emerald. doi: 10.1108/9781800715974

Park, M., Choi, Y., Yasui, M., Hedeker, D., & Specificity, Commonality, and Generalizability in Social-Emotional Development Special Section Editors. (2021). Racial discrimination and the moderating effects of racial and ethnic socialization on the mental health of Asian American youth. *Child Development, 92*(6), 2284–2298. doi:10.1111/cdev.13638

Perrin, A. (2016, February 18). English-speaking Asian Americans stand out for their technology use. Retrieved from *Pew Research Cente*r website. www.pewresearch.org/fact-tank/2016/02/18/english-speaking-asian-americans-stand-out-for-their-technology-use/

Ruiz, G. N., Edwards, K., & Lopez, H. M. (2022a, April 21). One-third of Asian Americans fear threats, physical attacks and most say violence against them is rising. Retrieved from *Pew Research Center* website. www.pewresearch.org/fact-tank/2021/04/21/one-third-of-asian-americans-fear-threats-physical-attacks-and-most-say-violence-against-them-is-rising/

Ruiz, G. N., Shao, S., & Shan, S. (2022b, August 2). What it means to be Asian in America. Retrieved from *Pew Research Center* website. www.pewresearch.org/race-ethnicity/2022/08/02/what-it-means-to-be-asian-in-america/?utm_source=Pew+Research+Center&utm_campaign=9e12ff243d-Weekly_2022_08_06&utm_medium=email&utm_term=0_3e953b9b70-9e12ff243d-400451109

Scott, J. (1988). Social network analysis. *Sociology, 22*(1), 109–127. doi:10.1177/003803858802200107

Smith, M., Rainey, L., Shneiderman, B., & Himelboim, I. (2014, February 20). Mapping Twitter topic networks: From polarized crowds to community clusters. Retrieved from *Pew Research Center* website. www.pewresearch.org/internet/2014/02/20/mapping-twitter-topic-networks-from-polarized-crowds-to-community-clusters/

Stieglitz, S., & Dang-Xuan, L. (2013). Social media and political communication: A social media analytics framework. *Social Network Analysis and Mining*, *3*(4), 1277–1291. doi:10.1007/s13278-012-0079-3

The Cybersmile Foundation. (2022). Online hate targeting Asian people spikes as coronavirus crisis deepens. Retrieved from The *Cybersmile Foundation* website. www.cybersmile.org/news/online-hate-targeting-asian-people-spikes-as-coronavirus-crisis-deepens

Watanabe, N. M., Kim, J., & Park, J. (2021). Social network analysis and domestic and international retailers: An investigation of social media networks of cosmetic brands. *Journal of Retailing and Consumer Services*, *58*, 102301. doi:10.1016/j.jretconser.2020.102301

Yang, C. C., Tsai, J. Y., & Pan, S. (2020). Discrimination and well-being among Asians/Asian Americans during COVID-19: The role of social media. *Cyberpsychology, Behavior, and Social Networking*, *23*(12), 865–870.

Ziems, C., He, B., Soni, S., & Kumar, S. (2020). Racism is a virus: Anti-Asian hate and counterhate in social media during the covid-19 crisis. *arXiv preprint arXiv:2005.12423*. doi:10.48550/arXiv.2005.12423

12 Mobilized Cultural Identities

Digital Friendship and Identity Maintenance among Japanese Immigrant Wives

Min Wha Han

Introduction

Extended from traditional computer-mediated communication (CMC), social messaging applications have transformed how people communicate. As LINE announces its app to be "transforming the way people communicate, closing distance between families, friends, and loved ones" ("LINE," 2022), social messaging applications in general, and LINE in particular, enable people to engage in synchronous communication using mobile devices (Russell, 2016; Chen, 2021). According to Chen (2021), one of the strengths of this social network application is its emotional expressivity, whereby it "has integrated multiple print-linguistic and non-linguistic forms of virtual aesthetic representation" (p. 237). Thus, its popularity can be explained not just by its discursive functions but also by its enabled visual aesthetics that allows closer connectivity among users.

Since its launching in 2011, subscribers for the app have kept increasing drastically, marking 218 million users worldwide in 2016 and 700 million in 2022. While originated and owned by a Korean online giant company, Naver, the largest and the most rapidly grown consumer market for LINE has been in Japan. It is evidenced by more than a 30 percent stock increase right after the 13 million shares offered to the Tokyo stock exchange in 2016 as a part of the Japan-US IPO (Rooney, 2016). Today, according to DMR Business Statistics (2022), the 78 million monthly active users of LINE in Japan mark the largest user population among a total of 230 countries, followed by Thailand (44 million), India (30 million), Indonesia (22 million) and Taiwan (21 million).

Numerous studies have been conducted on LINE such as its impacts on mobile phones and apps on health (Sato et al., 2017), the educational uses of mobile phone apps (Thornton& Houster, 2005), and most recently on the assessment of mobility change through an analysis of mobile data (Nagata et al., 2021). Yet, topics related to Japanese immigrant wives, as opposed to immigrant workers and youths, and the impact of mobile communication on their friendship maintenance and cultural identities have been largely overlooked.

This chapter explores the role of mobile apps in sustaining one's homeland culture and friendship, focusing on the use of the social communication app, LINE, among Japanese immigrant wives in the United States. Employing a netnographic

DOI: 10.4324/9781003328896-16

field research method and informed by the theories of adult friendship, cultural identity, and social presence, the study focuses on the use of the LINE app to understand how this social media app plays an important role among Japanese immigrant wives in their cultural group membership. Through qualitative content analysis and in-depth interviews, the chapter discusses communication and friendship characteristics that were exhibited in their LINE chat, and how those exhibited characteristics were supported by the features of the LINE app that enable the users to feel presence and connection. Paying attention to both content and relationship elements, the chapter examines how the app use operationalizes the members' cultural identities of *being* and *becoming*.

The study first contextualizes the topic within the literature of mobile communication, identity, and culture, leading to discussions of existing studies about the impacts of mobile communication on immigrant families. Further, to explain the uniqueness of the population for this study (middle-aged Japanese immigrant wives), discussions on adulthood friendship and the nature of cultural identities are followed by a discussion of media presence theory.

Mobile Communication, Identity, and Culture

Mobile technology has brought ubiquitous changes to modes of human relationships. Adaptation of social media and social network services in daily communication, relationship building, and maintenance has become increasingly common among people across the globe. If the rationale for making relationships, or our desire to form relationships across the lifespan for our own well-being has not changed, the ways in which people *approach* each relationship in their life course exhibit observable shifts due to technological advancement. In this rapidly shifting technoscape and its influence on human relationships, scholars have questioned how these trends impact the traditional notion of identity and culture, and how these developments intersect with negotiations of cultural identities (Anderson, 2019; Bell, 2005/2015; Bhandari, 2021; Ostendorf et al., 2022).

Cerra and James (2011) assert one of the prominent transformational trends is the power of social networking on people's connectivity, social engagement, and lifestyles. In their extensive study on the impact of technology on human identity, they analyze how this "networked community age" has brought a shift in the identities of teenagers and young adulthood. Young adulthood, coined by the term "emerging adulthood," and ranges from 18 to 25, Cerra and James call this age range as the life stage "in search of the ideal" (p. 113). According to Cerra and James (2011), in the networked community age, presence in social network services tends to play an important part in identities among people in this age. While their data mainly focus on Facebook, the argument remains intact as scholars increasingly observe the importance of social network service on individuals' identities (i.e., see Ostendorf et al, 2022).

The mobile phone in particular is seen as a main contributor to the networked age (Bell, 2005/2015; Geser, 2005). According to Geser (2005), the use of cell phones

> empowers and enlarges the sphere of micro-social interaction by making individuals free to reach each other under any circumstances and without the need to conform to institutional norms that demand a presence in a specific place (and relationship with others present at this same location).
>
> (p. 24)

Early studies of mobile communication from Asia confirm a high correlation between the use of cell phones and the strengthening of intimacy among close social ties (Park, 2003; Miyata, 2003).

This function of mobile communication that strengthens social ties, along with its ability to free up people from bounded place, brings scholars to an interesting argument regarding the notion of cultural identity. McCarthy et al. (2013) noted, we "live in a time of the radical separation of culture from place as globalizing forces and pressures comprise open local certainties, local forms of association, affiliation and feeling, local ways of dwelling in built space" (xvi). They observe that mobility as a fundamental condition of how the twenty-first century challenges traditional notion of cultural identity, in which "identity is seen as the true self within the collective self, *bounded by the material fixity of ancestry, language and place*" (xvi, emphasis added).

Theoretical Framework

This study adopts triangulated theoretical frameworks of cultural identity theory, adult friendships and social presence theory. Drawing from Bhandari's (2021) notion of cultural identity, Rawlins's notion of adult friendship, and social presence theory, the study looks into how LINE plays a central role in the formation of cultural identities and maintenance of group membership among Japanese immigrant wives.

Cultural Identities as Both Being and Becoming

As observed by Bhandari (2021), "cultural identity of an individual involves in dynamic processes of interaction with multiple factors" among which "social interaction and personal awareness play decisive roles" (p. 104). As such, the three conceptualizations categorized in Bhandari's (2021) review are (1) essentialism; (2) social constructivism; and (3) postulation of being and becoming, as posited by Stuart Hall. The first category, essentialism, assumes the predominant existence of inherited characteristics in an individual's cultural identity. This perspective asserts that the core cultural values serve as internalized "essential values" within an individual who is eternally transcending "time and space" (p. 105). Against the cultural deterministic position of the essentialist approach, the second position emphasizes the importance of the "social interaction of an individual for the formation of cultural identity" from a social constructivist perspective (p. 105). In the social constructivist's notion of social interaction, as Bhandari (2021) notes, "an individual may take up different cultural identities in different social and cultural contexts and their cultural identity undergoes a constant process of formation and reformation" (p. 105). Then Bhandari's third point, drawn from Stuart Hall's notion of cultural identity, recognizes both essentialist and social constructivist claims on identity. Cultural identity, especially that of immigrants, can be positioned at the intersection of being and becoming. That is, the concept of being involves "the basis of their shared cultural values and historical experience[,]" while the concept of becoming emphasizes "an ongoing process of production" in which self is recreated in the relational contingencies of the new environment (Bhandari, 2021, p.106).

When applied to the case of Japanese immigrant women, the states of both being and becoming play significant conceptual roles. In Japan's relatively homogenous and predominantly communal national culture, this notion of essentialism positions the self's cultural identity in the historically rooted origin, thereby setting the basis of common ground among the immigrants. Then, the immigrants' lives in their host country demand the self to be reconfigured. This notion of both being and becoming – the identity that is "contingent" yet "not ahistorical or immutable" (Bhandari, 2021, p. 106), is worth analyzing in mobile communication among the Japanese immigrant wives.

Adults' Identities, Friendship Characteristics, and Relational Dialectics

In his work on friendship, Rawlins (1992/2006) views adult friendship as broadly categorized into ages from and around 30 to 40 (times of maturity) as well as the 40s to around mid-60s (middle age). Adulthood friendships are characterized by a unique paradox of choices and constraints:

> As other periods of life, adulthood involves an ongoing configuring of self with others through communicating in diverse roles and relationships. As the adult years unfold, numerous choices regarding marital, parental, family, work, and community roles and performance define one's day-to-day possibilities and responsibilities while also restricting one's options for alternative life paths.
>
> (Rawlins, 1992/2006, p. 157)

One's adulthood life is simultaneously enabled and constricted by choices s/he makes previously in her/his life, creating "an increasingly complex array of contingencies" (Rawlins, 1992/2006, p. 157) that come after that. Importantly, as Rawlins (1992/2006) asserts, "each person's opportunities and obstacles are socially and culturally patterned" (p. 157). Gender, marital status, and socioeconomic status, for example, are equally important contributing factors as one's immigrant status. An important characteristic of adulthood friendship, therefore, is that friendship this time and in the future depends more on roles, relationships, and "communicative exigencies" (Rawlins, 1992/2006, p. 158) of each individual rather than chronological correlation or a direct tie to one's age.

Social Presence Theory

Rooted in multidisciplinary theories of social psychology, symbolic interactionism, and interpersonal communication, social presence theory today is widely used in media communication (Cui et al., 2013). Grounded in Wiener and Mehrabian's (1968) notion of immediacy and Argyle and Dean's (1965) concept of intimacy, social presence theory has gone through phases of evolution in terms of its contexts (e.g., face-to-face, telecommunication, CMC) and its complexity (from binary to multiple attributes). Today, in the field of media communication, social presence theory is concerned with various media attributes that contribute to the "feelings of presence" (Lee & Nass, 2005). As such, previous studies applied social presence theory to examine people's desires, experiences, and levels of satisfaction in online communication settings (Tu, 2001).

Central Research Questions

The proposed chapter addresses the following research questions:

RQ1: What characteristics of identity building as well as relationship maintenance strategies can be identified in Japanese immigrant wives' communication on LINE?

RQ2: How do the features of the LINE app support the exhibited characteristics and maintenance strategies?

Methodological Approach

Drawing from the triangulated theories of adult friendship, cultural identity, and social presence theory, this study employs a netnographic fieldwork (Kozinets, 1998) as a qualitative social media research method that includes observations, qualitative content analysis, and in-depth interviews. First introduced as a method in consumer market studies (Kozinets, 1998), netnography has been increasingly adopted as a both theoretical and practical viable research method (Boellstorff, et.al, 2012) that allows a researcher to access online interactions among community members to gain in-depth insights about them (Chao, 2015). While humanist participant observation remains a prominent research tool, netnography integrates online and offline interactions as field sites. In other words, it allows a researcher to understand humans with/using technology as well as humans within technology as sites of meaning-making. In this regard, technology is both a tool and a context.

For initial data collection and analysis, a netnographic participant observation was conducted on a group of Japanese immigrant wives who reside in a partly rural and partly suburban city in a southern state in the United States. The group consists of 18 members who are originally from Japan, ranging in age from their early thirties to late fifties. Most of them were married to American husbands, with a small number of exceptions for those who have Japanese husbands. The group's discursive and nondiscursive use of the LINE app was analyzed. Data presented in this article were based on the group chat conversations among the members during a period of 6 months between April and September 2022. Invited to be a member of the group, the researcher immersed herself in both online and offline interactions in the group during this time period. The online portions of conversations resulted in a total of 1,928 messages, including those of online stamps. Data analysis for this report is informed by principles of the constructivist grounded theory approach (Charmaz, 2006). In the constructivist approach the researcher "creates the data and ensuing analysis through interaction with the viewed [the researched]" (Charmaz, 2000, p. 523).

After getting approval from the Institutional Review Board, two focus groups and 10 one-on-one interviews were conducted. The focus group interviews along with one-on-one follow-up interviews were done to better explain and validate the textual data presented here, and further achieve data saturation (Charmaz, 2000, 2006; Corbin & Strauss, 2008). Focus group interviews took place in a conference room at a university library, and individual interviews were conducted online via Zoom. Each focus group interview lasted approximately three hours, while each individual interview lasted from 45 minutes to an hour.

Qualitative content analysis was applied to the discursive data categorizing each conversational thread into topic/content aspect, communication characteristics, and relational messages. Answers from focus group and individual interviews served to feed background information, including their history of mobile phone usage in Japan, family and cultural background, and marriage. Further, questions asked "what" and "how" of their LINE usages, such as the purpose (why they use it), and method (what major functions the app fulfills), along with their emotional attachment to this app. Data have been manually coded by the researcher and analyzed through the thematizations method.

Findings and Discussions

An analysis of the discursive and nondiscursive data along with interview data has revealed the prominent role the LINE app plays among Japanese immigrant wives in building cultural identities and maintaining their friendships. RQ 1 looked into characteristics of identity building as well as relationship maintenance strategies. Answering RQ 1, three prominent communicative characteristics found throughout the data analysis are: (1) a positive and animated tone; (2) voluntarism and inclusivity; and (3) member equality and informality. Further, while linguistic and cultural similarities served as a fundamental core for *being,* the roles they play and the supports they exchange with each other as immigrant wives create unique group identities of *becoming*. These characteristics contributed to their cultural identity of *being* Japanese, yet at the same time to the *becoming* of "families."

RQ 1: What characteristics of identity building as well as relationship maintenance strategies can be identified in Japanese immigrant wives' communication on the LINE app?

Positive/animated Tone, Voluntarism, and Inclusivity

Throughout the conversations, members sustained a positive and animated tone, which added inclusivity. Welcoming new a member, announcing an offline event, and celebrating each member's and her family member's birthdays were consistently made in a positive and animated tone. Adding new members to the group was always done manually with announcements to the members. For example, upon inviting a new person who had been introduced by a friend of one member, 82 messages were shared among members. These messages have characteristics that: (1) welcome the new member; (2) introduce existing members' backgrounds such as their homeland in Japan and the length of living in their current residing place; and (3) share information about family members.

Here, similar to Troilo and Britten's (2020) research on people's friendship maintenance strategies on Facebook, this part of the data confirmed positivity and a voluntary nature as a main characteristics to get a new member acquainted with the group members, although positivity alone does not necessarily guarantee the relationship's development into close friendship (Troilo & Britten, 2020).

In a group interview, members explained the importance of keeping a positive tone to encourage voluntary participation in conversations. One member explained this:

> It's totally ok to ignore or skip because everyone's lifestyle is different. … But then those who need this information, such as information with some urgency, and want to start the conversation would read and join.

This comment has led to a discussion of how "small talk" functions in their LINE conversations, and participants agreed strongly on its role to sustain open invitations to everyone:

> "So we think it's completely ok to be ignored. [Especially] like a story like [one member's] 'I played with my kids in a pigsty today!'"
> "Yeah, we have so many small talks in our posts, and it's totally ok to be skipped!"

Then in a personal interview, young members in their thirties tended to mention that they are "welcome and accepted" because of the positive and animated tone that runs through the messages. Members tend to understand that it is okay to "ignore" some messages, as is evidenced by a comment such as "I join conversations when timing meets, but when I am busy with my kids, I tend to skip;" or "During weekends or times with my family, I tend to silence the sound because the LINE counts each emoticon or every stamp as one message." Yet, they commonly mentioned that such "skipping" is accepted, and overall they feel "welcome and accepted."

Equality and Informality

Second, having diverged from the Japanese cultural norm of age-based hierarchy and politeness, members in this group strived to place equality and informality at the center of their friendship maintenance in the LINE conversations. These characteristics were kept by using nicknames throughout their online conversations and sometimes even sharing age-related jokes. Each member was called by a nickname rather than their real name. The use of nicknames then helps the members set the tone for their online interactions into informal rather than formal. For example, on several occasions, members "admonished" each other not to use honorifics that are norms in Japanese society among different age groups. This was initiated by elder group members (late forties to early and mid-fifties), who were in the group for longer periods of time rather than young and newer members.

A one-on-one interview revealed that this equality was initiated by some members in their forties and fifties who play a central role in the group. A woman in her fifties mentioned, "No *keigo* (honorific) here. It's okay to speak without it. As mothers, when you raise children, the age difference does not mean anything." Not surprisingly, then, when talking about the benefits of the friendship circle operationalized through the LINE app, this woman mentioned the group's multi age-group friendships:

> To me, the age difference is significant. So because we cannot meet and talk directly because of the physical distance, I think that the LINE covers this aspect. Like, if it was within a similar age group, we can actually meet and talk because we have something in common. But then if you are not in the same age group, perhaps we do not meet and talk much. I think it's the LINE app that filled the age gap.

A sense of equality and informality, then, seems to serve as one of the most effective relational maintenance strategies to sustaining their group identity.

Being Japanese, Becoming "Family"

Being Japanese, or from Japan, exists as the most fundamental sense of affiliation. A discursive analysis has revealed that the members of the Japanese immigrant wives group used the LINE to keep their emotional ties to Japan. The members shared various information related to Japan such as changing policies regarding the pandemic and the related visa issues. Like the first point, members also shared some critical moments that were both public and private. Private family information such as the loss of family members in Japan tended to be followed by enormous emotional support among members. Similarly, critical news from Japan was also shared with strong feelings. For example, when the former Japanese prime minister was assassinated in the summer of 2022, the news was shared and led to highly emotional conversations.

> "Oh my (shocked face emoticon)"
> "(Sharing a news article from Japan) He is in a state of cardiac arrest."
> "I want him to live."
> "I really want him to endure. On Japanese politics, I don't particularly support any politicians, but I don't want him to lose in this type of violence."
> "I can't believe that this thing (gun shooting) happens in Japan. It's sad to have this world."

Apart from their gender-role identities of mothers and wives, comments of this kind were underlined by shared ethnic and national identities. This position supports Anderson's (2019) observation on digital diaspora in that "the notion of home and one's relationship to the homeland hold a central position" and the "members of diasporic groups are considered to be linked together by a common interest in their location of origin and a foundational identity" (p. 154). In this regard, members of the Japanese immigrant wives presented a intense emotional connection to their homeland and the places of origin, albeit through an essentialized assumption that equalizes political and cultural identities.

A sense of connection due to the place of origin and linguistic proximity was also revealed in interviews. One of the youngest members, who grew up in Japan as a bilingual between Japanese and English, mentioned:

> I feel this group's members are emotionally closer than my other social groups. Because we all speak Japanese and we all have connections to Japan in some way, that's a special connection already made, and that's the core that we all have basically something in common. … Here in this group, I feel like we have reasons to connect. Even though there are people who I don't see often, emotionally I do feel close.

Supporting this statement, several members share how they were "lonely" before joining this group, in many cases due to language barriers they face living in the United States. Then, functioning as the fundamental membership condition, *being* Japanese and from Japan seem to validate shared standpoints among the Japanese immigrant wives.

While the sense of *being* plays a fundamental and a core role, the Japanese immigrant wives stressed a sense of *becoming* something that is beyond who they were in Japan. For example, sharing the roles of wives and mothers in a place that is foreign to them plays a critical function in maintaining their adulthood friendship circle. The breadth of the topics covers categories such as food, schools, stores, shopping sites, health, and weather. The members share pieces of information that are useful for other members such as where to find Japanese groceries, where to get medicines and vaccines, information on newly opening Asian restaurants, and forecasts of snowy weather or other severe weather forecasts. Then, in-depth conversations or expressions of deeper emotions were observed in some critical moments. For instance, when an active gun shooting tragedy occurred in Uvalde, Texas, the members shared lengthy conversations in the LINE's group chat:

> "This is extremely scary"
> "I am speechless … Why this kind of things can happen …"
> "As a parent who has an elementary schooler, really, so heartbroken."
> "Really, my heart is so broken."

This message exchange is followed by a lengthy discussion on teaching kids about gun safety.

"I believe we should teach kids about gun safety. But don't know how … (crying emoticon)."

"My husband told me that (in America,) even pre-K or Kindergarteners are taught (on gun safety) just like Japan's fire drill."

"We should also teach our kids not to touch it wrongly."

"True, this happened to my American uncle when ..."

As immigrant wives face the need to adapt themselves to the norms of the host country that is drastically different from their place of origin, this type of open conversation seems to serve as a vehicle for their relationship maintenance by confirming and validating their shared cultural backgrounds.

In interviews, members actively used the word "family" to describe their friendship circle. For example, supporting [each other] for "childcare" has been the most strongly agreed group identity as well as strategies to sustain close relationships.

"When (one member) was pregnant, we made support groups, for several different emergencies."

"Friends supported me with food even for a small operation after my delivery." "We can say…it's really a family." "Extended family for sure!"

The use of a "family" as a metaphor to describe their close-knit support was followed by the origin of this group. One member who is the longest member of the group mentioned:

> When I was in [another state], there was a support group among American wives. … I remember that I was really helped by them, so I wanted to do so to Japanese friends who are new here.

This routine of helping and being helped by members seems to contribute to the formation of a new identity, when one of the newer members, in her forties, mentioned, "People here are so warm-hearted like families, so I started to feel I also want to be helpful to them. … I think that's the part I see change in myself."

Importantly, the sense of *becoming* meant a divergence from communication *in* Japan. The Japanese immigrant members agreed with the notion of unique cultural codes shared only within this cultural group. One of the members, who is in her thirties, mentioned that this was the only Japanese group of people to whom she opened herself, explaining as follows:

> I don't think I feel close to the Japanese when I was in Japan. Both culturally and linguistically I feel outsider there. When I was a college student in Japan, I didn't feel comfortable (as a hafu Japanese identity), and at that time, I felt my identity as a Japanese had gone far. But then, since I met this group, I started to begin the Japanese parts in me, and even to appreciate them. I like to be with

> people who have both, and know how to appreciate both. That's *becoming* our new identity.

She further explained how the Japanese immigrant wives in this group are different from the Japanese whom she met in Japan:

> I feel like they understand both the good and the bad of Japanese culture, and this group tries to focus on the goods. And we even sometimes talk about the bads of Japan. That doesn't happen in Japan (laughter). [In Japan] you don't have that critical thinking ability that you get basically by coming to other countries and getting a different perspective. …Talking with people who know that makes me feel accepted.

The notion of creating a new culture that is *like* family, and one that is uniquely different from one's original culture/friendship circle in Japan, was consistently recognized through interviews.

The second research question tried to look into the nature of the LINE app's features that enable (and sometimes constrain) the above-discussed characteristics:

RQ 2: How do the features of the LINE app support the exhibited characteristics and maintenance strategies?

The findings reveal that characteristics of identity building, as well as the relationship maintenance strategies, were supported by the LINE's features that enable both immediacy and intimacy. A sense of immediacy was confirmed through the shared feeling of presence and having fun. The LINE app also helped the members to feel a sense of intimacy regardless the number of face-to-face interactions.

Immediacy: Enabling a Feeling of Presence and Fun

Interactions in the LINE app among Japanese immigrant wives were highly associated with presentations of emotional expressions. The positive and animated tone, voluntarism, and inclusivity were not only enabled by the textual tone and positive contents, but also enhanced by emoticons and virtual stickers. Through a discursive textual analysis, it became apparent that the groups' frequent use of emoticons and virtual stickers was one of their observable norms. In a focus group interview, the members agreed on the use of emoticons and virtual stickers as a way to add a feeling of presence:

> [With the use of emojis and stamps, we talk] as if we are having an ongoing teatime conversation. We are broad in an age spectrum, but we are so close-knit [due to this] … . Sometimes I feel like, it would be easier and faster to talk over the phone, but even at that moment, I kept talking/texting over the LINE. … I don't know why (laughter). Sometimes I feel it's strange.

In explaining the "strange" feeling, the speaker mentioned the joy of reading messages loaded with emoticons and virtual stickers. She mentioned that her talking over the phone tends to be limited to an occasion for a direct question or asking a favor, but reading messages in her relaxed position gives her joy.

Here, as confirmed by Chen's (2021) netnographical research on the use of the LINE app among young Taiwanese participants, the users actively employ the LINE's functions of virtual stickers "to make sense of other persons as well as build their self-image in terms of managing aesthetic identity and emotions in their LINE friendships" (p. 241). This thesis on added visual aesthetic functions to an online cultural identity was also apparent across the adulthood friendship circle among Japanese immigrant wives, perhaps due to similar cultural codes between the two nations. Importantly, the dominant characteristics of the virtual stickers used by the members of this group were convergent with Japanese female cultural codes to keep "in-group harmony and the 'cute' trends in taste" (Chen, p. 241). In this regard, the aesthetic identities formed and performed across online LINE messages presented rather "ideal" cultural selves with respect to the dominant Japanese gender and cultural codes, which were observed by the use of virtual stickers that present "playful emotions" (Chen, 2021, p. 245).

Intimacy: The LINE-initiated Relationship

Both focus group interviews and one-on-one interviews confirmed that the LINE enhances degrees of intimacy. Members used the LINE app for overall relationship maintenance involving Japanese, which included their friends and family in Japan. By using LINE, the members feel as if this app literarily "shortens the physical distance." At the same time, cues for psychological proximity have also been identified. A comment that supported this claim was this:

> Something that I think about this group is that … . [w]e cannot meet that often because we are all busy, but even those I have not met and talked to often, I feel like I knew that person *through the way sentences are formed or the "feel" from the message*.
>
> (Emphasis added)

This comment may suggest that discursive formations on the LINE app, be they sentences, emojis, a virtual sticker, or a combination of all, operates to supplement the personal attributes of the senders. It then would contribute to the knowledge of the person prior to the meeting as evidenced by other comments:

> "I feel like because of the LINE conversation we've already done [before meeting], the app makes it easier for me to become a friend, than meeting for the first time and start learning about them. It makes it easier to shorten the distance."
>
> "When we think about our actual meetings being so occasional, I think that the LINE is serving as a cushion, and to me, I feel like the relationship with others has come closer."

The above comments seem to confirm the social presence theory's contention for people's "desire [for a] high level of social presence and create environments promoting more social presence" (Cui, Lockee, and Meng, 2012, 676). The LINE-initiated self-disclosure, as a contributor to an enhanced social presence, seem to have a strong impact on the Japanese immigrant wives' adult friendships.

Conclusion

Given the active, vibrant, and frequent communications among these Japanese immigrant wives in LINE, presented findings and discussions add significant implications to the research of mobile communication and cultural identities. Particularly, the participants' ages range from early thirties to late fifties, the findings in this study are not limited to one particular age group, but rather offer answers to the research questions from a multi-age grouped perspective. As previous research on this topic was predominantly done on young generations (Chen, 2021), the fact that the data in this study covered *beyond* the young generation has significance. Furthermore, this study also leaves cultural implications as to the population outside the Western societies, an Asian immigrant community that has been under-represented.

One of the limitations of this study is its small data size in generalizing the findings. Although the sample of the discursive and nondiscursive messages was large in numbers, for a qualitative study more in-depth analyses of the nuances, variations, and reasoning behind the individuals' choices need to be addressed.

Nonetheless, the chapter hopefully addressed diverse yet coherent perspectives from the Japanese immigrant wives and the features of the LINE app that enable them to sustain their cultural identities and adult friendships. In closing the chapter, it might be worth sharing their comments on the "meaning" of this group. According to one of the youngest members,

> [w]ithout this group, perhaps I was losing the Japanese side of me. Japanese language and culture. Maybe I would lose chances to speak Japanese, and like foods, there would be no chance to eat if I don't cook it. Then I probably wouldn't! Because of the people in this group, I can hold on to this Japanese side. Without this group, I think I would be an American! ... I think that's not good. Because my home is Japan, but forgetting, or not feeling close to this root, is sad. I feel like I can hold on to my roots in a new way with this group.

Finally, one of the oldest members of the group contends this:

> If there was not this group, I think… .my life would have been boring. What matters, in the end, is … . it's being Japanese. No matter how long I live here, I cannot become an American in every aspect of my life. Maybe, because this group is here in my life, I can live here. … You know, there is [lots of] laughter in this group. We can laugh together.

References

Anderson, K. B. (2019). Digital diasporas: An overview of the research areas of migration and new media through a narrative literature review. *Human Technology, 15*(2), 142–180.

Argyle, M., & Dean, J. (1965). Eye-contact, distance and affiliation. *Sociometry, 28*(3), 289–304.

Bell, G. (2005/2015). The age of the thumb: A cultural reading of mobile technologies from Asia. In P. Glotz, S. Bertschi, & C. Locke (Eds.). *Thumb culture: The meaning of mobile phones for society*, pp. 67–88). Transcript Verlag.

Bhandari, N. B. (2021). Diaspora and cultural identity: A conceptual review. *Journal of Political Science, 21*, 100–108.

Boellstorff, T., Nardi, B., Pearce, C., & Taylor, T. L. (2012) *Ethnography and virtual worlds: A handbook of method.* Princeton University Press.

Cerra, A., & James, C. (2011). *Identity shift: Where identity meets technology in the networked-community age*. Wiley.

Chao, B. (2015). Relevance and adoption of netnography in determining consumer behavior patterns on the web. *The Scholedge International Journal of Business Policy and Governance, 2*(6), 12–17.

Charmaz, K. (2000). Grounded theory: Objectivist and constructivist methods. In N. K. Denzin & Y. S. Lincoln (Eds.), *Handbook of qualitative research, 2nd ed.* (pp. 509–535). Sage Publications.

Charmaz, K. (2006). *Constructing grounded theory: A practical guide through qualitative analysis.* Sage Publications.

Chen, C. (2021). Friendships through the style choice of virtual stickers: Young adults manage aesthetic identity and emotion on a social messaging line app. *International Journal of Market Research, 63*(2), 236–250.

Corbin, J., & Strauss, A. (2008). *Basics of qualitative research, 3rd ed.* Sage Publications.

Cui, Lockee, B., & Meng, C. (2012). Building modern online social presence: A review of social presence theory and its instructional design implications for future trends. *Education and Information Technologies*, *18*(4), 661–685. https://doi.org/10.1007/s10639-012-9192-1

DMFA (May 2020) Why LINE is the most popular social media app in Japan and how to use it for online ad campaigns. *Digital Media for Asia.* Retrieved from www.digitalmarketingforasia.com/why-line-is-the-most-popular-social-media-app-in-japan/

DMR Business Statistics (2022, September). DMR Line report. Retrieved from https://expandedramblings.com/index.php/line-statistics/

Geser, H. (2005/2015). Is the cell phone undermining the social order? Understanding mobile technology from a sociological perspective. In P. Glotz, S. Bertschi, & C. Locke (Eds.). *Thumb culture: The meaning of mobile phones for society* (pp. 23–36). Transcript Verlag.

Kozinets, R. V. (1998). On netnography: Initial reflections on consumer research investigations of cyber-culture. In A. W. Joseph & W. J. Hutchinson (eds.), *Advances in consumer research* (Vol. 25, pp. 366–371). Association for Consumer Research.

Kozinets, R. V. (2015). *Netnography: Redefined.* Sage Publications.

Laing, R. D. (1971). *Self and others.* Penguin.

Lee, K., & Nass, C. (2005). Social-psychological origins of feelings of presence: Creating social presence with machine-generated voices. *Media Psychology, 7*(1), 31–45.

LINE: Calls & messages. *Google Play.* Retrieved from https://play.google.com/store/apps/details?id=jp.naver.line.android&hl=en_US&gl=US

McCarthy, C., Kozma, A., Nicole L., Palma-Millanao, K., & Fitzpatrick M. (2013). Introduction. In K. Palma-Millanao, M. Fitzpatrick, & N. Lamars (eds.), *Mobilized identities: Mediated subjectivity and cultural crisis in the neoliberal area* (pp. xvi–xxvi). Common Ground.

Miyata, K. (2003, June 22–24). *The Mobile-izing Japanese: Connecting to the Internet by PC and Webphone in Yamanashi.* "Front Stage/Back Stage and the Renegotiation of the Social Sphere", Grimstad, Norway.

Nagata, S., Nakaya, T., Adachi, Y., Inamori, T., Nakamura, K., Arima, D., & Nishiura, H. (2021). Mobility change and COVID-19 in Japan: Mobile data analysis of locations of infection. *Journal of Epidemiology, 31*(6), 387–391.

Ostendorf, S., Meier, Y., & Brand, M. (2022). Self-disclosure on social networks: More than rational decision-making process. *Cyberpsychology, 16*(4), 1–27. doi: 10.5817/CP2022-4-2

Park, W. (2003, June 22–24). *Mobile phone addiction: A case study of Korean college students.* Front Stage/Back Stage and the Renegotiation of the Social Sphere. Grimstad, Norway.

Rawlins, W. K. (1992/2006). *Friendship matters: Communication, dialectics, and the life course*. Routledge.

Rooney, K. (2016, July 14). LINE spikes 30% in market debut after opening at $42 in largest tech IPO. *CNBC.* Retrieved from www.cnbc.com/2016/07/14/japanese-messaging-app-line-opens-at-42-in-largest-tech-ipo-of-the-year.html

Russell, J. (2016, June 12). *Understanding line, the chat app behind 2016's largest tech IPO.* Retrieved from https://techcrunch.com/2016/07/14/understanding-line-the-chat-app-behind-2016s-largest-tech-ipo/

Sato, Y., Kojimahara, N., & Yamaguchi, N. (2017). Analysis of mobile phone use among young patients with brain tumors in Japan. *Bioelectromagnetics, 38*(5), 349–355.

Sullivan, H. S. (1953). *The interpersonal theory of psychiatry.* Norton.

Thornton, P., & Houser, C. (2005). Using mobile phones in English education in Japan. *Journal of Computer Assisted Learning, 21*(3), 217–228.

Tu, C. (2001). How Chinese perceive social presence: An examination of interaction in online learning environment. *Educational Media International, 38*(1), 45–60.

Wiener, M., & Mehrabian, A. (1968). *Language within language: Immediacy, a channel in verbal communication.* Appleton.

13 "Respectfully, Pls Ask Someone Else"

Pride and Shame in International K-Pop Fandom

Samantha James

Introduction

Fans of Korean pop music (K-pop) have been increasing in the global limelight for their impressive feats of mass organizing. Whether renting a billboard in Times Square (BTS NYC, 2018) and in Seoul wrapping an entire train to celebrate an artist's birthday (Koreaboo, 2020), or matching a $1 million donation to the U.S. Black Lives Matter movement (Springman, 2020), the organizing power of these international collectives online is undeniable. However, when I conducted a series of preliminary interviews with self-identified super fans outside of Korea, I found that multiple participants felt pushed to withdraw from fandom after experiencing bullying or feeling ostracized from other online and offline social circles for their love of K-pop. These fans also stated that they only recently returned to fandom with the emergence of new social media platforms that allowed them both the anonymity to avoid shame and the ability to find others who share their passion. Inspired by interviewee responses, I began an ethnographic investigation into the communicative dynamics of K-pop fandom in Korea to see if fans sensed pride and shame the same way when performing fandom online and offline in Seoul, the heart of the K-pop industry. In the transition from American to Korean fan communities, I immediately noticed the deeply embedded role mobile communication technologies played in the design and function of K-pop fandom. This chapter illustrates my journey, as a U.S.-based K-pop fan and scholar, into the Korean fan community using autoethnographic inquiry to demonstrate how the fundamental role of mobile applications influences the sensation of pride and shame at different levels of the fan experience. First, I provide background on pride and shame in fandom, the role of mobile communication in global fandom historically, and the contemporary embodied role of mobile communication. Then I take an autoethnographic approach to illustrate how the embeddedness of mobile technology in K-pop fandom influences sensations of pride and shame within the community. To conclude, I discuss the implications of building a fan base enriched by mobile embeddedness in a contemporary globalized world.

DOI: 10.4324/9781003328896-17

Theoretical Background

Pride and Shame in Fandom

The experience of being a "superfan," that is, someone who dedicates time and effort to some form of content or celebrity, involves sensations of both intense pride and shame. Media production companies first coined the concept of "fandom" as a driver for revenue generation (Fraade-Blanar & Glazer, 2017), and fandom has since taken on a meaning all its own as fans of all forms of content created communities of practice around their object of fandom (Malik & Haidar, 2020). Fans organize to build community, to show support for their object of fandom, and even use the networks created to fight for social causes that seem beyond the scope of a traditional fandom (Kanozia & Ganghariya, 2021). Fans have a lot to be proud of, from impressive fundraising efforts for global social movements (Kanozia & Ganghariya, 2021) to simply creating spaces of inclusion, both online and offline (i.e., Fathallah, 2017; Jenkins, 2011; 2013; Kunert, 2021; McInroy, 2020).

However, fans still feel judgement and insecurity around their fan identity, partly due to the gendered history of pop culture fandom. The Western stereotype of a "silly fangirl" can be attributed originally to the time of "Beatlemania" (Ehrenreich et al., 2002). When international superstar boyband the Beatles first came to the United States, their arrival marked one of the first times the media, and therefore much of the public, saw vast swaths of screaming fans outside of an event venue. Since then, news media outlets began to describe fan groups as overzealous, dramatic, and at times dangerous, all stereotypes we still see today (Kussin, 2019; McGay, 2021). Fans cite "outing" themselves as fans to the communities they belong to beyond virtual fan spaces, often preferring to keep the two identities separate for fear of stereotyping. Despite the prevalence of fandom, fans of all types of media still face negative assumptions. In the case of K-pop, these negative stereotypes became highlighted and exacerbated with the global spread of fan networks online. Fans sense both pride and shame through their experiences in online and offline interactions with their favorite artists, non-fans, and themselves. Understanding how those sensations permeate hybrid contexts can teach us more about global organizing at large.

Global K-pop Fandom and Mobile Communication

K-pop is now inherently digital and increasingly global. What began as an opportunity for Korean popular music creators to receive funding from the government in the late twentieth century quickly became a global phenomenon, blooming in popularity in Southeast Asia and other East Asian countries. Networks of fans began to purchase albums and other merchandise in large quantities to save money on shipping abroad. In the early 2000s, K-pop began moving abroad, further west and into the Global South; and fan bases in South America and Central Asia blossomed alongside new media technology that allowed fans to easily gain digital access to their favorite content. By the time Psy's hit earworm track "Gangnam Style" took off, largely due to his viral music video (Gibson, 2018), many K-pop production companies were already targeting a North American market. Contemporary K-pop fans are inherently international in the way we trade physical and digital content across oceans and airwaves. As the Internet became more widespread, fans of K-pop, like many other subculture fandoms, shifted the means with which we communicate and share content around the world.

The K-pop industry's global popularity looks like a Chilean fan of heavy metal bands suddenly finding herself passionate enough about a K-pop group that she plans a vacation to Seoul just to see them perform in a festival. Fans like this one, whom I met at said festival, and many others, including myself, fell in love with K-pop first and reaped the benefits of a broadened cultural understanding after. I have connections to more countries than I ever had before and am better informed about current events and popular opinion in more places because of our shared online networks. The K-pop industry's global popularity also brings to light issues of race and gender in content and fandom reception. In the United States, for example, certain radio stations still refuse to play K-pop songs that are not in English despite their popularity. In Southeast Asia and South America, more conservative government leaders have disparaged the K-pop industry for leading fans to challenge traditional masculinity and family roles (Ainslie et al., 2018; Diaz Pino, 2021). Multiple western fans interviewed as part of this study cited instances of their peers making demeaning comments about the fan's favorite idol groups because of their Asian identity. As both a United States-educated academic and a K-pop fan, I place my own identity against the historical and contemporary atrocities committed in the United States toward immigrants and Asian Americans as well as recognize the still very racialized and gendered elements of contemporary fandom stereotypes around the world. As K-pop continues to globalize, we must understand this phenomenon through its digital embeddedness.

Digital Ubiquity and Embodiment

The expansive global K-pop fan network that exists today results from global marketing efforts by the industry, globalization, and increased media digitalization. During the first wave of COVID-19 lockdowns around the world in 2020, for example, production companies were able to capitalize on the K-pop industry's advanced technology and produce high-quality virtual concerts alongside the high-quality social media content they already relied on to reach a global audience from Seoul. In response, the industry gained fans at a time when live music was non-existent (Rapkin, 2021). While fans reported that these concerts were not as good as the "real thing" (Gibson & Herman, 2022), they still tuned in and paid to watch livestreams because of their dedication to the group, formed largely via digital networks. K-pop fans have been known to raise large sums of money online to donate to charities in an artist's name and strategically "stream" new songs on a variety of music platforms to raise the popularity of our favorite artists (Dooley & Lee, 2020). Most of the popular K-pop fan activities can happen anywhere in the world with Internet access. Even online, fans sense the strong emotional tug toward their object of fandom as they do in person. For K-pop fans, many of whom do not have the time, resources or the ability to see their favorite idols perform live, the online communities they participate in create similarly felt fan experiences.

The embeddedness of digital communication technologies in contemporary K-pop fandom is in line with contemporary Mobile Communication literature's claims that we are firmly in the mobile era. I agree with Jason Farman's (2020) claim that embodiment in the mobile era is sensory inscribed. While using social media to communicate with other fans, sensations of pride and shame permeate through online interaction and practices. Taking a social shaping approach to understanding Information and Communication Technologies, or ICTs, the study of practices and experiences on fan community platforms shows both how our world is shaped by technology and how in turn we shape the technology (Lievrouw & Livingstone, 2010). K-pop fans use digital communication technology to plan upcoming online and in-person events, react to online and offline events in real-time, and navigate shared identity, community, and cultural norms. Fans choose to participate from wherever and whenever they want, and only come together when they decide to participate. This eventful view of organizing (Ganesh & Wang, 2015) marks one of the intersections for the pride/shame dialectic to occur, as fans sense feelings of pride and/or shame throughout the experiences they choose to participate in online and offline. The interactive dialectic of pride and shame that surrounds the fandom experience permeates differently through global hybrid contexts. Emerging mobile communication technologies influence how we interact and sense emotions at an embodied level, and understanding the way that influence shifts sensations of pride and shame in K-pop fandom may also provide insight into those shifting sensations in other forms of global hybrid organizing. Therefore, I ask the following research question:

RQ: How does the embeddedness of mobile applications into the K-pop fan community influence sensations of pride and shame?

Methodological Background

This chapter is part of a larger ethnographic project surrounding K-pop fan communities in the United States and Korea. I began with in-depth interviews and hybrid ethnography of K-pop fandom events in the United States. From there, I moved to Seoul, home of the large production companies that created K-pop over two decades ago. Events in the United States pale in comparison to the intensity, size, and quantity of official (produced by the company) and unofficial (fan-led) events that take place in Seoul.

The transition to South Korea from the United States was a drastic one for more than just fannish reasons, and alongside the moments of frustration and exhilaration that surrounded novel experiences of moving to a new country came frequent recognition of the importance of mobile communication in the contemporary globalized world. I spent weeks navigating new banking, healthcare, and messaging applications required for immigration and integration into society. I also prepared to continue my ethnographic inquiry into K-pop fandom in Korea and found the integration of new mobile applications a requirement for participation in that community as well.

I love K-pop and acted as a fan, as well as a researcher, in the fan community. The process of *learning how to fan* in the heart of K-pop fandom inspired me to reflect on my transition and my role as an insider-outsider, and someone with a vested interest and knowledge of the fandom subculture who lacked much of the broader social knowledge around being a member of Korean society. Thus, reflection on my time as a long-term fan and new participant–observer in hybrid K-pop fandom in Korea provides meaningful insight into the sensations felt by other fans navigating the embeddedness of mobile applications in a hybrid global society.

Autoethnography as Method

To understand how the embeddedness of mobile applications into the K-pop fan community influences sensations of pride and shame, I utilized autoethnographic methods. Autoethnography is an individual's narrative situated within a culture (Adams et al., 2015), thus providing an understanding of the embeddedness of mobile communication technology into K-pop fandom in Korea from the ground up. According to Boylorn (2008), "Autoethnographers look in (at themselves) and out (at the world) connecting the personal to the cultural" (p. 413). I take my personal experiences navigating new systems of fan organizing, both formal and informal, within the broader context of Korean popular culture in the contemporary global sphere. My experiences are a mix of "Aha!" moments with mundane "aesthetic moments" (Adams et al., 2015) as well as my own frustrations, excitements, and reflections on the extremely interconnected state of networked fandom. To best situate my identity as an American fan and researcher within the larger conversation surrounding mobile communication in Asian communities, I provide an in-depth recounting of my experience learning to be a fan in Korea.

Building from a chapter written from a similar perspective (James, 2021), I use individual narrative as a lens to understand how my experience speaks to that of the individual versus collective sensations of hybrid international fandom. As a researcher as well as a fan, I do not enjoy K-pop in a vacuum. Context is vital to understanding cultural phenomena, especially music (Matula, 2000), and this chapter provides context and an in-depth analysis of the fragments that make up my personal experience in K-pop fandom to answer a larger question: What does mobile communication do in global fandom? Taking to heart this book's goal to understand mobile communication in Asian societies, I foreground the global nature of contemporary K-pop fan communities and study them from the heart of K-pop in Seoul. I do not intend to speak about Korean or Asian culture as a whole; rather, I use my experiences and accompanying questions, emotions, and reactions to theorize the effect that deep-rooted mobile connectedness may have on the specific subculture of K-pop fandom as it moves from online to offline.

I use self-reflexivity to situate my individual experiences as being both partial and illustrative of this cultural phenomenon. My experience as fan-turned-researcher includes the cultural fragmentation of experience that I am aware of bringing to the text. I begin with my primary introduction to K-pop fandom research, and briefly outline realizations and reflections from my time as a participant-observer and employee in a K-pop store in Austin, Texas, then recount my path of discovery as a new fan in Korea and share one of the first strong affective moments of pride and shame I felt viscerally through my mobile communicative practices. Next, I explore how different online practices shaped the in-person embodied sensations of pride and shame in a recent example of relearning fandom practices before turning to the discussion about themes that cut through both events.

Introduction to Fandom Research

I began my research inquiry into K-pop fandom by first interviewing fans in the United States alongside working myself at a K-pop store in Austin, where I watched fans grapple with their identities as fans of Korean popular culture *in person* for the first time. In spring 2022, I worked at a small family-run shop that specialized in Korean pop culture products. While we made most of our profit from selling stationery kits, pens, and other trinkets bought in bulk from overseas, most of the store's returning customers were K-pop fans coming to pick up the latest album to arrive from Korea. One customer, a young woman, said to me while checking out, "I can't believe I'm spending this much money on something so silly. I'm so delusional." This was the first time I was able to understand the sensations of fan pride and shame.

The time spent in the store as well as my ethnographic and interview work in the United States led me to naively believe I understood how K-pop fandom worked around the world. Fans I interviewed reported feelings of shame around the amount of dedication they had to K-pop. At this point, this seemed like a uniquely U.S. K-pop fan experience, potentially connected to the position K-pop played as "racial other" in the U.S. popular music industry. I had seen claims of "international versus Korean fans" online, but assumed that the fan base for K-pop was mobile-embedded such that that K-pop's role as a cultural counter-flow (Kang, 2023) from the East to the West was a backdrop for a new globalized fandom in which all fans acted the way I had seen on my social media, largely influenced by my English-language preferences, previous likes, and location settings. I began preparing to move to Korea to continue my research, hoping that certain fan communication practices would be enacted similarly to the United States. However, I quickly found that digital embeddedness does not detract from the localized physical experience, rather it shapes the experience within contemporary local context.

Autoethnographic Reflection: Embedded Mobile Communication in Korean Fandom

While using embedded mobile communication practices to learn how to be a "good fan" in different contexts, I have run into multiple powerful affective moments in moving from online to offline. I describe these examples as events, each one building off knowledge and understanding I had from the previous experience and included communicative elements from the prior reflective moment. After watching fans in the United States grapple with their offline fan identity as they moved from social media communities to in-person fan events, I moved to Seoul and quickly realized that fandom requires an entirely new set of mobile communication practices. I spent months learning the expectations for fan communities online and offline in a new context, and faced shifting sensations of pride and shame as I learned how much I still did not know. I draw from two specific affective events to explore how the digital embeddedness of K-pop fandom in Korea influences shifting sensations of pride and shame.

Event 1: "That's How it Has Always Worked"

As my research trip to Seoul neared, I began to question my understanding of fan organizing, so I turned to familiar online platforms and researched where non-Korean K-pop fans come together while in Korea. I found Facebook groups for fans specific to my local area in the United States and followed that model to find similar groups for non-Korean fans living in Korea. Those pages redirected to the popular Kakao messaging app, part of the Kakao media platform that is home to a nearly ubiquitous Korean money transfer app, directions app, public transportation app, and home shopping app (Yim, 2021). Already, I was encountering a new type of fandom shaped by culturally embedded social media practices.

Upon arriving in Korea, I learned that I had already missed ticketing for a concert I had been looking forward to attending. The first ticket-buying experience felt unreasonably difficult. A friend with prior experience told me that ticketing in Korea was easier than in the United States, sharing that most ticket-buying sites were quicker to navigate. However, I could only create an account on the global version of the website, despite being in-country, since I did not yet have a Korean address and could not pay on the website using a credit card since I did not yet have a Korean bank account or phone number. That was the first of many small roadblocks related specifically to K-pop fan activities in Korea. Even once I acquired an official identification card, bank account, mailing address, and phone number, there were certain fan-specific practices left to learn. I followed along in my Kakao chat groups specifically for international fans in Korea and saw fans share stories like mine. One fan complained that she was not allowed into an event because she had not brought all the right documentation; a laundry list of materials required by the fan club leaders, including government identification, a physical album, and proof of registered phone number as well as fan club membership. In response, another fan called out their complaints, saying "That's how it has always worked. It's your fault for not understanding the rules." This response highlighted a few contradictions about the small community of foreign K-pop fans living in Korea. Primarily, it challenged the assumption often made by non-fans that K-pop fandom is tightknit as well as constantly recruiting new participants. This interaction proved that certain fans did the exact opposite, actively gatekeeping their insider knowledge about fan club expectations. Further, it drove me to reflect on my identity as a white foreigner in Korea. If I wished to participate in fan events, I needed to learn how the system worked without placing the burden of translation or instruction on the Korean fan club leaders. If other non-Korean fans reacted to questions about participation the same way as the previous example, where should I seek help? This community of non-Korean fans of K-pop in Korea acted differently from the community I knew back home, despite the discernable differences between the two being only our physical location and our reliance on different mobile communication platforms. This distinction carried into my future interactions with the same community.

It was not until a few months after moving to Seoul that I successfully won a mobile lottery draw to attend an event like the one described above. This event was one of the first strong affective moments of pride and shame I felt viscerally through my mobile communicative practices. To complete all the steps required to properly participate as a fan-club audience member, I relied on assistance from other foreign-yet-expert fans that I built a relationship with outside of our online international fan communities. However, fan club leaders almost barred me from participation because my name on the official fan club app was different from that of my official Korean identification card (Samantha James instead of JAMES SAMANTHA). This cultural difference in name production, alongside the intense reliance on the fan club membership smartphone application as a source of proof, led me to feel alienated and lost the same way I did with many, often more serious, immigration issues that I had faced when moving to this new country. Despite K-pop fandom's "silly" reputation, I felt the shame of being unprepared to the same extent that I had a week prior at the bank and a month earlier while attempting to set up a phone plan. I felt like I did not belong and as if trying to fit into a community that I was not a part of. I felt guilty for not understanding the rules, although I had sought out answers from both the mobile communication platforms designed to inform fans about events and from fan friends I had made within and beyond my other online communities.

However, that shame was countered by the sense of accomplishment I felt when I made the cut, even just barely. When I entered the venue, a large stadium hosting a K-pop festival with a variety of acts, I felt my heart swell with pride as I realized our fan club group was the biggest. The group I was there to support was the headliner for the festival, and many of the other attendees were excited to see them perform. Since this event was meant to promote K-pop around the world and would be livestreamed online, the purpose of our fan club-allotted seats was to demonstrate the power of collective K-pop fandom. I was so excited to do just that, screaming the rehearsed chants and shaking our glowing light sticks along to the beat with the rest of the fans while my favorite K-pop group took to the stage. The effort needed to participate in this event as an official fan club member heightened my feelings of pride. I was suddenly proud of myself for navigating this complicated process, proud of my favorite artists for being so popular, and proud of the other fan club members for demonstrating how to be "good fans" as we showed off how much we loved our group. The whiplash feelings of pride and shame from navigating a new fan culture carried through to my other fan events.

Event 2: "Respectfully, Pls Ask Someone Else"

After months of living and acting as a fan in Seoul, I still frequently learn new K-pop fan procedures. I recently attended an official fan club event for an up-and-coming K-pop group that is managed by a smaller production company and therefore uses an entirely different system of event registration and participation. To attend the desired event, I first signed up for a (free) official fan club membership on a specific mobile application. Then I watched for an announcement post that included a link to a Google Form, a deadline for submission, and instructions for how to participate. While written in English, in their vagueness these instructions mimicked those in the applications I used previously. At the exact time of application, my friend and I hit *submit* on our forms as quickly as possible, as only the first 100 people to submit to the event application were able to attend. Two weeks and six attempts later, the company announced an event with a higher acceptance rate, and we were allowed to attend.

Unlike the large festival previously described, this event was for a small number of fans to watch the group record a performance of their newly released single. This style of event is common in Seoul and for more popular groups very competitive to attend, and I was grateful for the chance to attend one. Despite my knowledge of other processes like this one, however, I was confused by the items we were required to bring to this event. Since the group's albums sold out, fan club leaders allowed an alternative option of showing proof of digital download. When we arrived at the venue – a recording studio about an hour outside downtown Seoul by public transportation – we were told by production company staff to line up in order of application submission number (shared to us via a spreadsheet when winners were announced online), and then asked to show proof of online fan club membership via the cell phone app, official Korean identification card, and physical printout copies of the group's most recent song on the downloads page of a Korean music streaming platform. I now know that this practice was one frequently used by K-pop fandom managers five to ten years ago, but the printing of a physical page of my online downloads to show as proof when I held a mobile application with the same information felt so unreasonable that, at first, I did not believe the instructions were accurate.

Given our status as new fans of the group, my friend turned to social media to ask a more experienced fan what exactly we needed to do for the fan club managers to allow us to attend the event. That fan, who had a high following count on Twitter and posted about this group in English quite frequently, responded positively to the first few messages we sent, despite having never met my friend before. However, when asked for confirmation that it was "normal" to print out a physical copy of the song downloads page, they responded quite harshly, referring to our question as equivalent to expecting them to act as our secretary. "Respectfully," they wrote, "Pls ask someone else." I was so ashamed. We asked another friend, whom we know more personally, with over fifteen years of K-pop fan experience, and she responded much more openly to our question, having last used a similar process nearly a decade prior. When we saw the rude fan in-person at the event the next day, only recognizable by her status as one of the few other non-Korean fans, we avoided her in embarrassment and frustration.

However, much like my other fan club experiences, I was overwhelmed with pride once the event began. When we watched the stage performance, I was so proud of both my friend and myself for navigating this difficult system, as well as of the artists for their talent at a young age. In the weeks following the stage performance recording, I have seen this group grow in popularity on platforms such as TikTok and YouTube. Videos of the performance we watched and others like it have gone viral, and their song is trending on social media. I am so proud to call myself a fan of this group, who are working hard to be recognized as global artists, and I am excited to say I got to see them while they were just starting out, even though we hit many small roadblocks in attempting to see them perform. Even after months of learning fan practices in Korea, navigating a slightly new system of online practices affected my physically embodied sensations of pride and shame. Breaking into those virtual and physical spaces has been difficult. Attempting to build relationships with fans as individuals or collectives has been as well, demonstrating how mobile communication influences connection as well as sensations of pride and shame in this fan subculture.

Discussion

The events described above illustrate how I felt pride and shame throughout my hybrid experience of learning how to be a fan in Korea through online and offline communication. Illustrating sensational "events" of transitioning from U.S.-based to Korean K-pop fandom elucidated multiple drivers for shifting pride and shame. In the stories described above, the sensations of urgency around event applications, alongside their inability to explain in full, lead to a reliance on the fan community. The digital embeddedness of official fan club content does not lessen inter-fan communication but rather shifts it to a hybrid form of digital-physical interaction in which fans are constantly doing both. Next, I describe how this hybrid embeddedness increased sensations of *time urgency*, *physical proximity*, and *weaponized knowledge* that drove my whiplash feelings of pride and shame as a non-Korean K-pop fan living in Korea.

Primarily, the feeling of *time urgency* related to being a member of the fandom – especially regarding mobile communication methods – shifted feelings of pride and shame. In the events described above, fans sensed a time urgency in the connection between mobile applications and event attendance. To attend both the large festival and the performance recording, I had to submit a form as close to the minute as possible and then hope I had submitted fast enough. Winning submissions were milliseconds apart at most. The adrenaline that comes from participating in these speed-based raffles is unmatched, and the dread that comes from failing them is embarrassing. When I was accepted into events, I felt extremely proud of myself for pulling it off, as well as a bit ashamed that I put so much effort into the process.

When I failed the raffles, a much more common outcome, I was simultaneously frustrated with myself for not being fast enough and ashamed that I care so much about the outcome when I know this is not a fair fight. Within the community, fans become competitors instead of collaborators. Despite the fandom's well-documented cases of showing extreme organizational and collective action skills, my experience changed when time was introduced as a factor of the K-pop fan experience.

Second, the *physical proximity* between me and my favorite idols changes feelings of pride and shame. K-pop was born in Seoul, and almost all active K-pop idols live in the city. As the center of fandom, it is easier for an artist to host more fan meeting events or for their production company to sponsor an official pop-up merchandise store for fans to attend. This influences the connection fans of K-pop have when physical proximity is much more common. Both events I describe above included the opportunity to see our favorite artists perform live. Fans fear shame from non-fans who may consider them as fitting the stereotypes created by news media since the time of Beatlemania. If a K-pop fan does not know the rules for participation in a live show – that is, which items to prepare for entry – there is a chance that fan does not know the unspoken rules of behavior during the show. If a fan makes us "look bad," by not acting the way other fans deem acceptable, then the stereotypes might carry on. The shame I felt when I was almost unprepared to enter the music festival was more than just a feeling of being uninformed: I felt that I had let my favorite artists down. To avoid this shame the next time, I reached out to friends with experience whom I met in other networks to ensure that I had prepared fully. The embeddedness of mobile applications with vague instructions plays into the requirement of fans learning, on their own, the rules of participation.

Finally, the transition from U.S.-based to Seoul-based fandom has changed the way fans compete for and *weaponize knowledge*. Before moving to Korea, I joined a series of fan groups on different social media platforms to share information. As seen in the above story of moving from online to in-person fan events, the process of attending an in-person fan event in Korea is shaped first, and throughout, by multiple mobile actions, from signing up for a membership in time to even knowing where to check for event announcements in the first place. K-pop production companies are frequently at the forefront of new social technology adoption because K-pop fans are willing to change processes frequently. Fans who have been around through various process transitions, especially those who have remained in Korea for multiple years, are often the gatekeepers of the necessary information that cannot be found by only using an Internet search or by following the rules in official announcements. Certain fans actively withhold information necessary for accessing events to limit competition, despite the generally assumed purpose of fandom to be building up the fan base. Knowing who, when, and how to ask people for help is an aspect of being a part of the fan community, defining who counts as a "good fan" in a group with no real defined boundaries or rules. Fans share in sensations of pride, and they shame one another to uphold expectations that benefit the fandom.

Implications and Future Directions

The embeddedness of mobile communication in Korea's K-pop fan community heightens sensations of time urgency, physical proximity, and competitive knowledge. In turn, these sensations cultivate a unique duality of pride and shame in each fan. I personally feel conflicting sensations of guilt and pride when participating in fan activities in Korea. I exhaust my financial, social, and mental reserves to learn all that I can about "how to be a good fan," using all the mobile applications (and the networks they afford me) at my disposal. I reach out to all my connections and ask for advice on how best to win a lottery-style chance to see my favorite artists perform live, while knowing that I am competing with those same connections for limited slots in the event. In turn, my dedication to fandom has increased tenfold since committing to this style of fandom. The elation felt when I see my name on a list of audience members is worth the hours that I researched, the money I spent, and the anxiety I felt while applying for a seat. My internalized shame breeds collective pride when I find other fans who behave just as passionately as I do, and thus we create a powerful collective of extremely dedicated followers.

From a practical application as well as a critical fan-studies perspective, future research should pay close attention to who benefits the most from the deep interconnection of mobile technologies and fandom sensations of pride and shame. Companies may benefit, but certain fan club members also earn social status (and sometimes financial gain) from their working knowledge and access to the platform of choice. The sense of urgency created by companies and fostered by fans also functions almost as a form of gambling, as fans compete with one another to see who can get as close to their favorite idols as possible while maintaining their role as acceptable members of the fandom. Future research should investigate the invisible boundaries between "good" fan and "bad" fan, and who benefits from the creation and maintenance of those boundaries.

Taking a mobile and global communication approach, future investigations into other contemporary contexts of mass mobile organizing should take K-pop fandom to heart. Further research on K-pop fandoms specifically should ask questions concerning identity differences within fandom and the impact of globalization and digital ubiquity on fan communities. The role of pride and shame in event-driven organizing such as in the case of K-pop fandom also informs us about issues surrounding other contemporary digital collectives. Social movements, political campaigns, and even hate organizations are shaped by some of the same factors that make up current K-pop fandom and, thus, studies of fandom should be taken to heart even beyond fan studies communities. Finally, this phenomenon is illustrative of global communication practices. Future research should investigate how mobile embeddedness affects, and is affected by, physical location settings, man-made algorithms, and other hybrid mobile technologies.

In conclusion, in this chapter I pulled from moments of embodied sensation to elucidate how the embeddedness of mobile communication technologies increased factors that shaped pride and shame in the context of global and local K-pop fandom. I illustrated how heightened sensations of time urgency, physical proximity, and competitive knowledge change the way fans embody pride in shame in a community where physical and mobile experiences intersect. This chapter illustrates my journey as a U.S.-based K-pop fan into the Korean fan community and used autoethnographic inquiry to demonstrate how the fundamental role of mobile applications influences the sensation of pride and shame at different levels of the fan experience. When building a fan base enriched by mobile embeddedness in a contemporary globalized world, producers must attune to the way fans embody pride in shame in a community where physical and virtual cultures intersect.

References

Adams, T. E., Jones, S. L. H., & Ellis, C. (2015). *Autoethnography: Understanding qualitative research*. Oxford University Press.

Ainslie, M. J., Lipura, S. J., & Lim, J. (2018). Understanding the potential for a Hallyu "backlash" in Southeast Asia: A case study of consumers in Thailand, Malaysia, and Philippines. *Kritika Kultura,* (28), 63–91. http://dx.doi.org/10.13185/KK2017.02805

Boylorn, R. M. (2008). As seen on TV: An autoethnographic reflection on race and reality television. *Critical Studies in Media Communication*, *25*(4), 413–433. https://doi.org/10.1080/15295030802327758

BTS NYC [@btsnewyorkcity]. (2018, October 12). Jimin Birthday billboard in Times Square, NYC! Thank you for this @parkjamjam_kr! It plays once an hour, between:00 and:10 past the hour. Located at 43rd and Broadway. #HAPPYJIMINDAY #JiminOurSerendipityDay #JiminOurAngelOurWorld #지민_널만나고_사소한건없었다 #BTSNYC https://t.co/opLPZ4YrNE [Tweet]. Twitter.

Diaz Pino, C. (2021). "K-pop is rupturing Chilean society": Fighting with globalized objects in localized conflicts. *Communication, Culture and Critique*, *14*(4), 551–567. https://doi.org/10.1093/ccc/tcab047

Dooley, B. & Lee, S-H. (2020, October 14) BTS's loyal army of fans is the secret weapon behind a $4 Billion valuation. *The New York Times*. www.nytimes. com/2020/10/14/business/bts-ipo.html

Ehrenreich, B. Hess, E., & Jacobs, G. (2002). 5. Beatlemania: Girls just want to have fun. In L. A. Lewis (Ed.) *The Adoring Audience: Fan culture and popular media*. Routledge.

Farman, J. (2020). Mobile interface theory (2nd ed.). Taylor and Francis. Retrieved from www.perlego.com/book/1683848/mobile-interface-theory-embodied-space-and-locative-media-pdf

Fathallah, J. (2017). *Fanfiction and the Author: How fanfic changes popular cultural texts.* Amsterdam University Press.

Fraade-Blanar, Z., & Glazer, A. M. (2017). *Superfandom: How our obsessions are changing what we buy and who we are*. Profile Books.

Ganesh, S., & Wang, Y. (2015). An eventful view of organizations. *Communication Research and Practice*, *1*, 1–13. https://doi.org/10.1080/22041451.2015.1110290

Gibson, J. (2018). How K-pop broke the West: An analysis of Western media coverage from 2009 to 2019. *International Journal of Korean Studies*, *12*(2), 22–47.

Gibson, J., & Herman, T. (2022) Entertainment and equity in the era of "ontact." In Sojin Lim (Ed.) *South Korean Popular Culture in the Global Context* (pp. 60–76). Routledge.

James, S. (2021). NeoCulture Technology: Affective fandom and K-pop in America. In D. Kim (Ed.) *Diffusion of Korean Popular Culture in Western Countries*. (pp. 63–80). Seoul National University Press.

Jenkins, H. (2011). "Cultural acupuncture": Fan activism and the Harry Potter Alliance. *Transformative Works and Cultures, 10*. https://doi.org/10.3983/twc.2012.0305

Jenkins, H. (2013). *Textual Poachers: Television fans and participatory culture* (Updated 20th anniversary ed). Routledge.

Kang, J. (2023). The politics of Being a K-pop fan: Korean fandom and the "Cancel the Japan Tour" protest. *International Journal of Communication, 17*(9). https://ijoc.org/index.php/ijoc/article/view/18885

Kanozia, R., & Ganghariya, G. (2021). More than K-pop fans: BTS fandom and activism amid COVID-19 outbreak. *Media Asia, 48*(4), 338–345. https://doi.org/10.1080/01296612.2021.1944542

Koreaboo (2020, August 30). BTS Jungkook's Fans Wrapped an Entire KTX Train In Birthday Messages – Here's How That Looks. Koreaboo. www.koreaboo.com/news/bts-jungkook-fans-wrapped-entire-ktx-train-birthday-messages-china/

Kunert, J. (2021). The footy girls of Tumblr: How women found their niche in the online football fandom. *Communication & Sport, 9*(2), 243–263. https://doi.org/10.1177/2167479519860075

Kussin, Z. (2019, May 28) *How BTS mania spawned a toxic K-pop fanbase that attacked me online. New York Post*. https://nypost.com/2019/05/18/how-bts-mania-spawned-a-toxic- k-pop-fanbase-that-attacked-me-online/

Lievrouw, L. A., & Livingstone, S. (2010). *Handbook of New Media: Social shaping and social consequences of ICTs, Updated student edition*. SAGE Publications. https://dx.doi.org/10.4135/9781446211304

Malik, Z. & Haidar, S. (2020): Online community development through social interaction – K-Pop stan twitter as a community of practice, *Interactive Learning Environments, 2020*, 1–19. https://doi.org/10.1080/10494820.2020.1805773

Matula, T. (2000). Contextualizing musical rhetoric: A critical reading of the pixies' "Rock music." *Communication Studies, 51*(3), 218–237. https://doi.org/10.1080/10510970009388521

McGay, E. (2021, March 1). *K-pop Fandom: Toxicity incarnate*. UD Review. https://udreview.com/k-pop-fandom-toxicity-incarnate/

McInroy, L. B. (2020). Building connections and slaying basilisks: Fostering support, resilience, and positive adjustment for sexual and gender minority youth in online fandom communities. *Information, Communication & Society, 23*(13), 1874–1891. https://doi.org/10.1080/1369118X.2019.1623902

Rapkin, M. (2021, July 7). *COVID-19 couldn't stop K-pop's global rise*. National Geographic. www.nationalgeographic.com/travel/article/covid-19-couldnt-stop-kpop-global-rise

Springman, L. (2020, August 3). *"It's ARMY versus the U.S. Army": K-Pop Fans, Activism, and #BlackLivesMatter*. Flow. www.flowjournal.org/2020/08/its-army- versus-the-army/

Yim, H-s. (2021, September 8). *South Korea ranks fifth in social app spending as KakaoTalk and YouTube take lead. The Korea Herald*. www.koreaherald.com/view.php?ud=20210908000773

14 Future Outlook of Mobile Communication in Asia and Research Directions

Ming Xie

Introduction

Mobile communication is the most rapidly expanding communication technology in the world (Gergen, 2008). The 12 studies in this book have explored the usage and implications of mobile communications in various spheres and contexts, such as Bangladesh, China, India, Japan, and South Korea. In this concluding chapter, we summarize and highlight the significant arguments indicated by these studies and propose our suggestion for future research direction regarding mobile communication in the context of Asian culture and society.

Cultural and Social Factors in Asia Intersecting with Mobile Communication

Asian society has been going through a Westernization process of urbanization and individualization. The process has had a profound influence on every aspect of society and social life, including parenting, romantic relationships, business culture, and other social norms. Traditionally, Chinese families are large, with several generations living together. However, with the accelerated urbanization and economic development process, the number of empty nesters has been rising. The study by Yang, Lachlan, and Chen (Chapter 2) connects mobile communication and the psychosocial well-being of empty-nest parents in such a context in China. With the significant increase in mobile communication usage, studies on mobile communication's function in reducing empty nesters' loneliness and depression and providing social support will have significant implications in addressing mental health and aging issues in Asian society.

Cultural values are also reflected by gender perceptions. With the interplay of feminism and traditional Chinese culture that views women as subordinate to men, gender roles and gender identities become more complicated. Xie and Chao (Chapter 3) explored the gender schema and feminism in Chinese society and how the role of housewives has been perceived and portrayed in the digital environment. From the perspective of digital feminism, they asked a critical question regarding whether social media and mobile communication simply replicate cultural and mass-media normative discourses or provide a more diverse and inclusive

DOI: 10.4324/9781003328896-18

perspective regarding gender roles in contemporary society. Digital technology creates new forms of feminist expression and feminist culture (Chang & Tian, 2021). In addition to the existing studies on how digital media promote gender equality, more attention can be paid to the potential of digital media to mobilize the feminist-networked public and social resistance regarding cultural tradition and political participation.

In the digital world and age of globalization, cultural norms and traditions have been significantly impacted by the advancement of technology. Cultural identity is another aspect of cultural norms and values. With the increasing migration of the global population, cultural identity has become fluid and diverse. Information technology and mobile media also prompt individuals to construct and negotiate their identities. Han (Chapter 12) studies Japanese immigrant wives in the United States and discusses how their use of mobile communication shapes and influences their cultural identities and relationship building. In contrast, the study of James (Chapter 13) describes a US fan of Korean pop culture and her use of mobile communication to help her navigate Korean culture and society.

Lastly, the Westernization process in Asia happens not only in the private sphere and family lives but also in the organizational setting and the business sector. With the increasing prevalence of neoliberalism and consumerism, e-commerce and m-commerce, especially livestream e-commerce have grown exponentially in Asian countries, including China. Yang (Chapter 6) discussed livestream e-commerce and its impact on social norms and business culture in China. The growth of e-commerce and m-commerce will remain to be the major trends in the international and global economy. Consumers are able to look for and find the product, share reviews with other consumers, make online payments, and complete the transaction process without stepping out of their homes. During the 2023 US Super Bowl game, a commercial advertisement for Temu attracted the audience's attention. Temu, is a US shopping app launched and owned by the Chinese e-commerce company Pinduoduo (拼多多). Like its Chinese version, Pinduoduo, Temu provides heavily discounted and low-priced products. As of February 2023, Temu was ranked second on the Apple App Store and first on the Google Play store (Kharpal, 2023). In addition to Temu, there is also another Chinese online shopping platform, Shein, that is accessible in the United States. In the digital world, consumers can choose a variety of online platforms to select, compare, and purchase goods and services (Dumanska et al., 2021). The emerging platforms, such as Shein and Temu, are worthy cases to study mobile communication, consumer behavior, and business models in the context of international business and economic development.

Mobile Communication across Spheres and Sectors

Crossing and bridging spheres and sectors are significant characteristics of mobile communication. Numerous theories and studies have highlighted the main function of mobile communication to build an expanded and horizontal network beyond boundaries (Fernández-Ardèvol & Ribera-Fumaz, 2022). It has to be noted that, although the book is divided into four parts – the private sphere, the organizational sphere, the public sphere, and the networked society – these spheres are intertwined with each other and cannot be viewed exclusively and separately, especially within the context of mobile communication. For example, mental health and public health issues happen across the private, organizational, and public spheres. In many countries, because of the cultural traditions, insufficiency of public service and social resources, and economic disparity, mental health and public health always face critical challenges to meet the needs of the public.

Several studies in this book examine the role of mobile communication in addressing mental health and public health issues in Bangladesh, India, and China. Dey and Averbeck (Chapter 4) compared Facebook's private groups and public pages and explored the at-risk population's mental health and their use of mobile communication regarding self-disclosure and seeking social support in different spaces in Bangladesh. Within the context of India, with 60–70 million people affected by mental illness, Subramanian (Chapter 7) studied the use of mobile communication by non-governmental organizations in mental health care. Built upon the well-known framework of nonprofits' social media engagement – information-community-action (Lovejoy & Saxton, 2012) – this study revealed how non-governmental organizations in India adopted destigmatizing strategies of education, contact, and protest through mobile communication.

Similarly, public health issues can be explored from both the perspectives of private and public spheres. Also focusing on the context of India, Kabir and Ha (Chapter 8) compared users' behaviors on different smartphone operating systems in health promotion during the COVID-19 pandemic. Focusing on the public health campaign combating COVID-19, their research identifies different communicative styles associated with different operating systems. In contrast, Li, Zhang, and Zhu (Chapter 10) explore WeChat users' health information-seeking activities, especially the COVID vaccination information, during the COVID-19 pandemic in China. Their research found that smartphones and mobile apps may be key players in narrowing the digital gap among users with different socio-economic statuses.

Digital Democracy in Asia

With the proliferation of information technology and social media, the discussion of the public sphere has expanded from mass media to social media. Discussions have been around whether technologies can create a new environment for democratic institutions that are more participatory and citizen-centric (Accenture, 2009; Tapscott et al., 2008; United Nations E-Government Survey 2010, 2010). Asian countries have diverse political cultures and institutions, ranging from authoritarian regimes and transitional countries to democratic and liberal states. A question that has drawn attention is whether and how mobile communication contributes to political life and the democratic process in authoritarian and transitional regimes (Gergen, 2008). Ding (Chapter 5) studied disaster management in the authoritarian regime of China and explored the role of mobile communication as a co-production instrument across public, the community, and governmental agencies in flood management. He argued that mobile communication had empowered citizens in various domains and promoted citizen participation and self-organized emergency management. In non-democratic and transitional regimes, the responsibility of providing public goods and public service is always viewed as taken by the state. The participation of non-state actors can be viewed both positively and negatively. On the one hand, non-state actors can take part of the burden of financial and administrative resources; on the other hand, non-state actors might be viewed as political competitors who challenge the state's authority and legitimacy (Tsai, 2011). The studies on China and other Asian countries provide valuable insight into the state–society relationship and the role of mobile communication in civil society and public-goods provision.

Compared to authoritarian China, Bangladesh's political system has been identified as semi-authoritarian, hybrid, or partly free, although it is a parliamentary representative democratic republic (Rahman, 2022). The study of Sharma (Chapter 10) explores the interplay of online and offline activism in Bangladesh and discusses how mobile communication forms a new public sphere based on a distributed network. With the demand to extend the limits of parliamentary democracy and allow citizens to express dissent, the movement organizers adopted both traditional and online mobilization tools to share information and motivate participants. This research resonates with previous studies regarding the function of mobile communication in initiating dialogues and interactions between movement organizers and participants and, therefore, fosters openness, trustworthiness, and commitment (Kruse et al., 2018; Loader & Mercea, 2011; Park & Reber, 2008).

Social movement studies have discussed online social movements such as #BlackLivesMatter, #MeToo, and Stop Asian Hate (Bonilla & Tillery, 2020; Lyu et al., 2021; Olteanu et al., 2015). These movements offer opportunities to address the stereotypical perpetuation of minority communities and institutional inequality and explore the contribution of social media to the successful mobilizations of social movements. Focusing on the #VeryAsian movement, Guo and Lipschultz (Chapter 11) conducted a social network analysis on social media content fighting against anti-Asian hate speech and crimes. On the one hand, social media spread anti-Asian sentiment during the pandemic; on the other hand, social media has also become an important tool for Asians and Asian Americans to produce counterspeech and fight against hateful sentiment.

Overall, mobile communication is a critical component of the information infrastructure in contemporary society. For countries with low socio-economic status and limited access to information technologies, mobile communication might play a significant role in community collaboration, social movements, and civil society development. Questions are raised regarding whether mobile communication platforms are alternative structures of public spheres by encouraging broader participation within various social and political contexts.

Future Research

Mobile Communication in the Context of Globalization and Localization

Based on the insightful and innovative discussion above regarding mobile communication in various contexts, we propose several suggestions regarding future directions for research on this topic. First, a comparative and contextual perspective is needed to develop theories of mobile communication based on practices within the local context of Asian society and communities. Reviewing the theories and theoretical frameworks adopted in the studies in this book, communication theories such as uses and gratifications theory, gender schema theory, social support theory, and affordance theory were applied in analyzing mobile communication practices in Asia that contribute to the applicability of these theories in a non-Western/non-Eurocentric context. In addition to those common theories, the rich and diverse cultures and contexts of Asia also provide opportunities for scholars to develop theories based on the localized evidence. The culturally specific insights and idiosyncratic perspectives will contribute to the global conversation regarding mobile communication and its social impact.

Second, scholars should pay more attention to the localization of mobile communication in various Asian countries. In this book, scholars examined mobile communication platforms such as Facebook, Instagram, LINE, TikTok, Twitter, and WeChat. Although most of the platforms are accessible all over the world, platforms such as LINE and WeChat are dominant social media platforms in countries such as China, Japan, and South Korea. Also, like Temu and Shein mentioned above, TikTok, owned by a Chinese company and gained global popularity, has been subject to criticism.

The localization of mobile communication reflects the social, cultural, economic, and political contexts of a country and a society. "The systems, arrangements, and values in which platforms are immersed shape platform logics and effects" (Sherman & Siravo, 2020, p. 1). For example, the different mobile platform's policies on privacy, user data, and information protection are associated with state regulation and culture. Also, the operation and management of mobile communication platforms might prioritize the owner companies' profit-seeking purposes. Jin (2013) analyzed the most used Internet platforms and found that most of them were run by for-profit organizations and utilized the targeted advertising capital business model. Moreover, the divergent social affordances of those platforms may facilitate different modes and infrastructures of communication and thus have distinct implications for the public sphere. We suggest scholars recognize the contextual uniqueness of sociopolitical factors to study the mobile communication platforms that have not been extensively studied yet.

From a Networked Society to a Mobilized Society

Technological innovation and the prevalence of smart and mobile platforms and devices have prompted scholars to reimagine the context, processes, and scale of human communication (Yao & Ling, 2020). Castells (2000) suggests that in a network society social, economic, political, cultural, and technological transformations are produced and reproduced through digital interactions. Compared to computers and laptops, portable and mobile devices allow users to be connected anywhere and anytime. The unprecedented mobility of connections has profoundly changed how people navigate their everyday social life (Campbell, 2020). On the one hand, people have the opportunity to connect with others beyond physical and geographical boundaries; on the other hand, scholars have been concerned that mobile communication might limit individuals within their small and intimate networks (Campbell, 2020; Kobayashi & Boase, 2014). As Matsuda (2005) argued, mobile communication might "facilitate an insular life with little attention to the public and the other" (p. 139). The simultaneous bridging and bonding effect of mobile communication, in addition to the evolving variety of mobile communication platforms, makes human communication much more complicated. Scholars face both the opportunities and challenges to explore how mobile communication advances people's mobility and social connections, and therefore fosters a networked public sphere at the local and global levels.

Overall, information technology is embedded in social, cultural, and political contexts. The 12 studies in this book cannot present a full picture of the application of mobile communication in diverse Asian communities. However, the collection offers concrete footholds and helps advance a conversation regarding mobile communication in relation to every aspect of our social life. Future studies on regional, cultural, and socio-economic nuances of mobile communication are needed to address the challenges and issues we face as we enter a new era of artificial intelligence.

References

Accenture. (2009). *Web 2.0 and the next generation of public service* (p. 40). http://apo.org.au/system/files/17610/apo-nid17610-15106.pdf

Bonilla, T., & Tillery, A. B. (2020). Which identity frames boost support for and mobilization in the #BlackLivesMatter movement? An experimental test. *American Political Science Review*, *114*(4), 947–962. https://doi.org/10.1017/S0003055420000544

Campbell, S. W. (2020). Cutting the cord: Social and scholarly revolutions as CMC goes mobile. *Journal of Computer-Mediated Communication*, *25*(1), 101–110. https://doi.org/10.1093/jcmc/zmz021

Castells, M. (2000). *The rise of the network society: The information age: economy, society and culture*. Wiley.

Chang, J., & Tian, H. (2021). Girl power in boy love: Yaoi, online female counterculture, and digital feminism in China. *Feminist Media Studies*, *21*(4), 604–620. https://doi.org/10.1080/14680777.2020.1803942

Dumanska, I., Hrytsyna, L., Kharun, O., & Matviiets, O. (2021). E-commerce and m-commerce as global trends of international trade caused by the Covid-19 pandemic. *WSEAS Transactions on Environment and Development*, *17*, 386–397.

Fernández-Ardèvol, M., & Ribera-Fumaz, R. (2022). The network society today. *American Behavioral Scientist*, online first. https://doi.org/10.1177/00027642221092800

Gergen, K. (2008). Mobile communication and the transformation of the democratic process. In J. E. Katz (Ed.), *The handbook of mobile communication studies* (pp. 297–310). MIT Press

Jin, D. Y. (2013). The construction of platform imperialism in the globalization era. *TripleC: Communication, Capitalism & Critique. Open Access Journal for a Global Sustainable Information Society*, *11*(1), 145–172. https://doi.org/10.31269/triplec.v11i1.458

Kharpal, A. (2023, February 13). *Chinese e-commerce giant PDD splashes on Super Bowl ad for its Temu U.S. shopping site*. CNBC. www.cnbc.com/2023/02/13/super-bowl-2023-temu-ads-launched-by-chinese-e-commerce-giant-pinduoduo.html

Kobayashi, T., & Boase, J. (2014). Tele-cocooning: Mobile texting and social scope. *Journal of Computer-Mediated Communication*, *19*(3), 681–694. https://doi.org/10.1111/jcc4.12064

Kruse, L. M., Norris, D. R., & Flinchum, J. R. (2018). Social media as a public sphere? Politics on social media. *The Sociological Quarterly*, *59*(1), 62–84. https://doi.org/10.1080/00380253.2017.1383143

Loader, B. D., & Mercea, D. (2011). Networking democracy? Social media innovations and participatory politics. *Information, Communication & Society*, *14*(6), 757–769. https://doi.org/10.1080/1369118X.2011.592648

Lovejoy, K., & Saxton, G. D. (2012). Information, community, and action: How nonprofit organizations use social media. *Journal of Computer-Mediated Communication*, *17*(3), 337–353. https://doi.org/10.1111/j.1083-6101.2012.01576.x

Lyu, H., Fan, Y., Xiong, Z., Komisarchik, M., & Luo, J. (2021). Understanding public opinion toward the #StopAsianHate movement and the relation with racially motivated hate crimes in the US. *IEEE Transactions on Computational Social Systems*, 1–12. https://doi.org/10.1109/TCSS.2021.3136858

Matsuda, M. (2005). Mobile communication and selective sociality. In M. Ito & M. Matsuda (Eds.), *Personal, Portable, Pedestrian: Mobile phones in Japanese life* (pp. 123–142). MIT Press.

Olteanu, A., Weber, I., & Gatica-Perez, D. (2015). Characterizing the demographics behind the #BlackLivesMatter movement. *ArXiv:1512.05671 [Cs]*. http://arxiv.org/abs/1512.05671

Park, H., & Reber, B. H. (2008). Relationship building and the use of Web sites: How Fortune 500 corporations use their Web sites to build relationships. *Public Relations Review*, *34*(4), 409–411. https://doi.org/10.1016/j.pubrev.2008.06.006

Rahman, T. (2022, September 6). *From Revolutionaries to Visionless Parties: Leftist politics in Bangladesh*. Carnegie Endowment for International Peace. https://carnegieendowment.org/2022/09/06/from-revolutionaries-to-visionless-parties-leftist-politics-in-bangladesh-pub-87806

Sherman, S., & Siravo, J. (2020). *We Don't Yet Fully Know What Platforms Can Do: Ten points towards public platformation*. Autonomy. https://autonomy.work/wp-content/uploads/2020/10/FINAL-Platforms-We-Need-Platforms-We-Want.pdf

Tapscott, D., Williams, A. D., & Herman, D. (2008). *Transforming government and governance for the twenty-first century* (p. 24). nGenera Corporation.

Tsai, L. L. (2011). Friends or foes? Nonstate public goods providers and local state authorities in nondemocratic and transitional systems. *Studies in Comparative International Development*, *46*(1), 46–69. https://doi.org/10.1007/s12116-010-9078-4

United Nations e-government survey 2010: Leveraging e-government at a time of financial and economic crisis. (2010). United Nations.

Yao, M. Z., & Ling, R. (2020). "What is computer-mediated communication?" – An introduction to the special issue. *Journal of Computer-Mediated Communication*, *25*(1), 4–8. https://doi.org/10.1093/jcmc/zmz027

Index